Complete
BUILDING
EQUIPMENT
MAINTENANCE
Desk Book
SECOND EDITION
Supplement

Edited by

SHELDON J. FUCHS, P.E.

PRENTICE HALL
Englewood Cliffs, New Jersey 07632

Prentice-Hall International (UK) Limited, *London*
Prentice-Hall of Australia Pty. Limited, *Sydney*
Prentice-Hall Canada, Inc., *Toronto*
Prentice-Hall Hispanoamericana, S.A., *Mexico*
Prentice-Hall of India Private Limited, *New Delhi*
Prentice-Hall of Japan, Inc., *Tokyo*
Simon & Schuster Asia Pte. Ltd., *Singapore*
Editora Prentice-Hall do Brasil, Ltda., *Rio de Janeiro*

10 9 8 7 6 5 4 3 2

Disclaimer

The techniques and methods presented in this book are the result of
the authors' experiences in working with certain materials and tools.
The information contained in this book is of broad general usefulness.
The reader should not use the information in this book without first
having a thorough understanding of his or her particular equipment,
environment, and organization, and care should be taken to use the
proper materials and tools as advised by the authors. The information
contained in this book is as up-to-date as it can be and has been carefully checked by the authors. Nevertheless, it is difficult to ensure that
all the information given is entirely accurate for all circumstances. The
Publisher disclaims liability for loss or damage incurred as a result of
the use and application of any of the contents of this book.

Library of Congress Cataloging-in-Publication Data

Complete building equipment maintenance desk book. Supplement
    edited by Sheldon J. Fuchs. — 2nd ed.
        p.   cm.
    Includes index.
    ISBN 0-13-157553-8
        1. Buildings—Mechanical equipment—Maintenance and repair.
    2. Plant maintenance. I. Fuchs, Sheldon J.
    TH6013.C66   1993 Suppl.
    696—dc20   93-20739
                                                                  CIP

ISBN 0-13-157553-8

PRENTICE HALL
Career & Personal Development
Englewood Cliffs, NJ 07632

Simon & Schuster, A Paramount Communications Company

Printed in the United States of America

DEDICATION

This book is dedicated to the memory of my mother, father, and wife, Myrna, whose love, devotion, inspiration and dedication made this undertaking possible.

And also

To the eight people in my life who make waking up in the morning have meaning and dreams still possible—Linda, Laura, Ron, and Dave and my second team—Julie, Allison, Michael, and Jacqueline.

And also

To Sylvia, a very special person who has a special part in my life and makes it worthwhile.

About the Editor

Sheldon J. Fuchs, P.E., is an associate with CFW Associates in Merrick, N.Y. His background includes Plant and Facilities Manager for the Baldwin School District, Superintendent of Buildings and Grounds, Director of the Facility Maintenance Institute and adjunct professor at Hofstra University, and plant maintenance engineer at Kollsman Instrument Corporation.

He holds a bachelor of civil engineering and a master's degree in business administration from the City College of New York. He has continued his graduate study at New York University and Hofstra University and is a licensed professional engineer in New York State and a certified plant engineer.

His articles have appeared in many trade journals and he has been a frequent speaker at seminars and meetings of various professional engineering societies. He is a contributing author to the *Plant Engineers Manual and Guide* and the *Encyclopedia of Professional Management.* As past president of the New York and Metropolitan Manhattan chapters of the American Institute of Plant Engineers (AIPE), he has been active in various professional organizations. Mr. Fuchs is founder of the Hofstra University Annual Plant Maintenance Seminar and Exhibit and has been awarded the Certificate of Merit from Nassau County (State of New York). He is currently a consultant with the State University of New York's College of Technology at Farmingdale, assisting them regarding curriculum, possible cooperative programs with industry and various start-up tasks regarding its new degree program in facility technology. In addition, he will be coordinating various continuing education programs for individuals who are working in the facility management, plant engineering, and related fields.

Contributing Authors

John A. Bernaden is a manager for the Controls Group of Johnson Controls, Inc., headquartered in Milwaukee, Wisconsin. In addition, he is Chairman of the Intelligent Buildings Institute's marketing committee in Washington, D.C. Bernaden co-authored the *Intelligent Buildings Sourcebook* in 1988 and *Open Protocols: Communication Standard for Building Automation Systems* in 1989, distributed by Prentice-Hall, Englewood Cliffs, New Jersey.

For the past three years, Bernaden has been the speaker for IFMA's two-day "New Technologies in Building Systems and Controls" seminar presented in locations across the United States.

He has also spoken about intelligent buildings, open protocols, and new building technologies at the AIA Congress, the IBI Congress, Intellibuild Conferences, World Energy Engineering Congress, Facilities '89, HVAC & Building Systems Congress, Northwest Energy Connections, and the Open Protocols Symposiums.

Bernaden joined Johnson Controls in 1983 and most recently has been a team leader in the design and development of the revolutionary Personal Environments concepts. He is a graduate of Marquette University in Milwaukee.

Christopher Branton holds a B.S. in biochemistry from Juniata College and an M.S. in Industrial Hygiene from Temple University. He has extensive experience in the major areas of industrial hygiene including, personal monitoring, ventilation controls, protective equipment, noise control, and training as they pertain to the manufacturing industry. In addition, Mr. Branton has considerable experience in the environmental field including, spill control and cleanup, hazardous waste disposal, remediation technologies, and soil and water sampling. His experience in both of these areas has allowed him to present a practical approach to safety, health, and environmental issues.

Salvatore T. Cordaro is a graduate of New York State Maritime College and holds a degree in chemical engineering from Brooklyn

Polytechnic Institute. He is president of Cord Associates, a management consulting firm. Mr. Cordaro is an experienced manager, having worked 30 years in the process industries. As a consultant, he has served many major corporations designing and implementing successful maintenance policies and programs. He is in great demand nationally as a seminar lecturer presenting regular maintenance seminars for North Carolina State University and Fairleigh Dickinson University.

Evans J. Lizardos, P.E., is a principal of Lizardos Engineering Associates, P.C., a consulting engineering firm that specializes in heating, ventilating, air conditioning, electrical, plumbing, fire protection, and energy conservation for commercial, institutional, municipal, and industrial projects. Mr. Lizardos holds a B.M.E. from Polytechnic University (formerly Brooklyn Polytechnic Institute), is a fellow of ASHRAE, and is a member of ISA. He has been heavily involved in the conceptual planning and implementation of automatic temperature controls.

Robert Nigrello earned a bachelor of engineering degree in electrical engineering from Manhattan College. He worked for the consulting engineering firm Stone and Webster where he did many site lighting projects. He is currently the North East Regional Specifications Manager for Osram Sylvania lighting. Mr. Nigrello is responsible for all liaisons with utilities, lighting designers, and energy service companies, introducing new products, applying new technology and troubleshooting lighting systems. He has earned the title of Certified Energy Manager (CEM) and Certified Lighting Energy Professional (CLEP). He is active in the Illuminating Engineering Society and is a board member of the New York IES section and a member of the National IES Merchandise lighting committee.

Paul D. Tomlingson is a veteran of 23 years of worldwide maintenance management consulting, the author of 3 textbooks and numerous trade journal articles on maintenance management. Tomlingson is a 1953 graduate of West Point, and in addition to a B.S. in engineering, holds an M.A. in government and an M.B.A., both from the University of New Hampshire. He is listed in *Who's Who in the West.*

If You Have Maintenance Responsibility This Supplement Is for You

The six contributors to this first supplement to *Complete Building Maintenance Equipment Desk Book, Second Edition* are outstanding specialists who are well versed not only in theory but also in the how-to aspect of plant engineering and maintenance. They have pooled their talents to produce a practical information guide that will be used as a constant source of information by those who are involved in all aspects of maintenance.

This supplement presents six chapters packed with ideas, checklists, guides, maintenance procedures, and concepts that will enable you to improve your operation and get the maximum from every dollar spent.

Provided are proven ideas and techniques that can double, triple, or quadruple profits—resulting from implementing a moderate, cost-effective equipment maintenance program. Every idea, every method has been fired in the furnace of real world application. This cornucopia of practical answers offers the best thinking of a cadre of experts in the field, people who have been faced with the same problems you confront and found workable, manageable solutions. Collectively, the cost-saving, equipment-saving, labor-saving examples from which they tap a rich bedrock of experience, have boosted the bottom line of actual companies by hundreds of millions of dollars.

The first step of improvement is evaluation—to establish current performance, identify what needs to be improved and the priorities. Equally important is the confirmation of what's being done well and telling people about it.

In chapter 1, Paul Tomlingson tells how to evaluate maintenance while getting full cooperation from those being evaluated; successfully converting the results into a plan of action and then, getting total, willing assistance to ensure lasting improvements.

Chapter 2 covers key points about the purchase and installation of a computerized maintenance management system. Based on personal experience, Sal Cordaro attempts to put logic in the justification of computer applications for maintenance. He discusses potential cost savings and principal benefits to expect. Also discussed are major activities to include in a maintenance software system. Cordaro highlights the common pitfalls that the new system user is subject to and tells how to avoid them. He includes 12 clarifying flowcharts to show application relationships in a PC-based maintenance management software system.

In chapter 3, Bob Nigrello provides the maintenance managers with details for optimally maintaining and troubleshooting their present lighting system. Incandescent, Fluorescent, and High Intensity Discharge Systems are covered. Lighting retrofits are discussed to achieve maximum kilowatt reduction, increased lamp life, and improve color rendering without sacrificing light output. Retrofitting case studies are provided to show real life energy savings. Current legislation on energy codes, lamps, and ballasts is covered in detail.

Chapter 4 discusses how automating the building equipment will maximize comfort for building occupants while minimizing the amount of energy utilized. The term "intelligent building" has become popular to define the state-of-the-art building automation system. John A. Bernaden covers items such as setting and controlling building thermostats, methods of seasonal changeovers, determining the breakeven temperature, making adjustments for the heating and the cooling seasons, the flywheel effect, and air handling systems design.

In chapter 5, Christopher Branton provides an overview of the major safety, health, and environmental issues that may affect your facility. The chapter begins with a review of basic chemistry principles, including corrosive and flammable materials, as they apply to the safety and health of employees. Safety issues covered in this chapter include personal protective equipment, confined spaces, exposure monitoring, and indoor air quality. Environmental issues include underground storage tanks, PCBs, asbestos, hazardous waste disposal, and spill procedures.

In chapter 6, Evans J. Lizardos discusses several maintenance procedures, detailed for both large tonnage and smaller tonnage refrigeration systems. These include purge control, refrigerant pump-out, relief pressure devices, venting, and certification. Many of these maintenance procedures are at issue when facilities managers are considering the service life of existing CFC-based equipment, equipment retrofits to switch to the newer refrigerants, or the purchase of new complete systems.

Sheldon J. Fuchs

Acknowledgments

This author is proud to be associated with individuals who have dedicated themselves not only to be professionals on their job, but to act as professionals when they leave their place of business. These individuals—whether their title is plant engineer, facilities maintenance manager, superintendent of buildings and grounds, public works administrator, or whatever—provide the leadership to keep businesses, schools, universities, hospitals, and everything else functional. In short, without these dedicated individuals, who are generally on call 24 hours a day, 7 days a week, 52 weeks a year, society would be "up the creek."

In addition, the various societies that represent these individuals have demonstrated to me that they do meet the needs of this profession and should be complimented for their value and efforts to improve this most important field.

The societies that I have had the honor to work with and have proven to me that they are truly professional are the following:

American Hotel and Motel Association

American Institute of Plant Engineers

American Institute of Public Works

American Society of Heating, Refrigeration, and Air Conditioning Engineers

American Society for Hospital Engineers

American Society of Mechanical Engineers (Plant Engineering Division)

Association of Physical Plant Administrators of Colleges & Universities

Building Owners and Managers Institute International, Inc.

International Facility Management Association

National Association of Power Engineers

Superintendents of Buildings & Grounds in New York State

The contributing authors, who themselves are experts in their fields, express their gratitude to all those who have generously assisted in the preparation of this supplement and have made a significant contribution to the text. These include, among others:

Dan Agne, Johnson Controls, Inc.

Norma Blake, Osram Sylvania

Gordon D. Duffy, Publishing Director, Engineered Systems

Jim Fox, Air Exchange Inc. 1185 San Mateo Ave, San Bruno, Ca. 94066

Gregg Homeyer, Raytheon Company

George Huhnke, Johnson Controls, Inc.

John McKewen, Osram Sylvania

Vincent Monaco, Avis Rent-a-Car

Cynthia Newton, Johnson Controls, Inc.

Glen L. Ponczak, Johnson Controls, Inc.

Harold Schneider, P.E., Office of Management & Budget, City of N.Y.

Dr. Roy Takekawa, Director, Environmental Health and Safety, University of Hawaii

Professor John Tiedemann, Chairperson for the State University of New York's College of Technology at Farmingdale

Table of Contents

Why There Are Inadequate Maintenance Evaluations (2)
A 10-Step Strategy for Conducting Successful Maintenance Evaluations (3)
    Step 1—Develop a Policy for Evaluations (3)
    Step 2—Provide Advance Notification of the Evaluation (4)
    Step 3—Educate Personnel on the Purpose of the Evaluation (4)
    Step 4—Schedule the Maintenance Evaluation (5)
    Step 5—Publicize the Content of the Maintenance Evaluation (6)
    Step 6—Use the Most Appropriate Evaluation Technique (8)
    Step 7—Announce the Results of the Maintenance Evaluation (10)
    Step 8—Take Immediate Action on the Results (11)
    Step 9—Announce Specific Gains Resulting from the Maintenance
      Evaluation (12)
    Step 10—Specify the Dates of the Next Maintenance Evaluation (12)
How to Use Evaluation Results to Improve Maintenance (12)
    Improving Supervisor Training (12)
    Monitoring and Improving Craft Training (14)
    Defining the Maintenance Program (14)
    Developing a Maintenance Mission Statement (14)
    Establishing Useful Maintenance Policies (16)
    Standardizing Maintenance Terminology (16)
    How to Establish the Maintenance Workload (17)
    Measuring the Productivity of Maintenance Workers (18)
    Conducting an Effective Preventive Maintenance Program (19)
    Establishing More Effective Maintenance Planning (19)
    Establishing a Responsive Work Order System (20)
    Ensuring Accurate Field Labor Data (21)

1

How to Evaluate the
Maintenance Function
to Improve Performance

Paul D. Tomlingson

As a maintenance manager, you know that quality performance by your department helps assure plant profitability. Conversely, when maintenance performance is poor, costs will be excessive and unnecessary downtime will cripple plant efforts to meet production targets and stay within operating budgets.

Unfortunately, desired performance improvements are often not realized because the evaluation of maintenance—the first step of improvement—is not done adequately. Too often, it is not done at all.

Perhaps you have observed at your plant that production gets a lot of attention. Their performance is evaluated frequently, in detail and usually very effectively. You may wonder why the same attention is not given to maintenance. Often it is because the same plant managers who insist on evaluating production operations may fail to realize that maintenance performance requires the same emphasis. Without evaluations, the specific improvements in maintenance performance that could assure profitability are often never identified.

While you may be keen on evaluations, you have, no doubt, acknowledged that support for maintenance from production, purchasing, and so on, has significant influence on maintenance performance. Unfortunately, you have little control over how much support is provided or its quality.

This chapter provides guidelines for how to conduct effective evaluations and suggestions for how to assure better support from those departments over which you have little control.

WHY THERE ARE INADEQUATE MAINTENANCE EVALUATIONS

A primary reason for poor evaluations is misunderstanding. This leads to suspicion and causes resistance. Typical:

- Maintenance may hope to delay an evaluation by creating meaningless "performance data."
- Warnings of unpleasantness with the union may be suggested to create uncertainty.

Under such circumstances, you may not go ahead with the evaluation. Thus, problems that should be corrected are never identified.

Maintenance resists evaluations because there are too many factors over which they have no control that influence their performance. Typical:

- The warehouse may be poorly run because it is managed by a remote, disinterested accounting department.
- Equipment may be unreliable because it is being pushed beyond maintenance due dates by operators who are pursuing unrealistic production targets.

Resistance must be eliminated by removing any threat to maintenance implied by an evaluation. Therefore, the evaluation must distinguish between those factors over which maintenance has responsibility and those over which it has no control—like the unresponsive warehouse run by an indifferent accounting department. Responsibilities for performance over each group of factors must be clearly established. Thus, if poor maintenance performance is due to poor warehouse support, this must be brought out into the open and corrected. You can evaluate maintenance fairly only if you also evaluate such aspects as operations cooperation, staff support and management sanction.

Maintenance, after all, is a service. Your staff cannot compel anyone to comply with your program. Your staff needs support and cooperation to be successful. Once it is clear that the factors over which you have little control will also be evaluated, most resistance to evaluations will disappear.

Maintenance also resists evaluations because of the limited management background of many of its key people. Maintenance managers rarely come from outside of maintenance. Rather, they are usually inbred and carry their management styles from the foreman to superintendent level unchanged. Many are graduates of the "school of hard knocks." University degrees, not usually a requirement, are rare among maintenance leaders when compared with operations, for example. Few have had any management training. They assume their supervisory responsibilities with little preparation for the challenges they will face. Some may even be unwilling supervisors and uncomfortable in the role. The majority are ex-craft personnel. Although they are competent in a single skill, they are often unfamiliar with the management techniques necessary to meet their obligations as supervisors. Further compounding the situation is the fact that a large number, as ex-craftsmen, are former union employees who may harbor an "us versus management" viewpoint.

A 10-STEP STRATEGY FOR CONDUCTING SUCCESSFUL MAINTENANCE EVALUATIONS

To take advantage of the improvement that maintenance evaluations offer, an overall strategy is necessary. The strategy must identify actions to organize and conduct an evaluation and convert them into improvements. A 10-step strategy includes the following points:

1–Develop a policy for evaluations

2–Provide advance notification

3–Educate personnel

4–Schedule the evaluation

5–Publicize the content of the evaluation

6–Use the most appropriate evaluation technique

7–Announce results

8–Take immediate action on results

9–Announce specific gains

10–Specify dates of the next evaluation

This strategy, which is described and illustrated below, recognizes that maintenance has the potential of contributing significantly to profitability. Maintenance will not evaluate itself. You must acknowledge that to some maintenance departments an evaluation is like a self-inflicted wound. You must be the catalyst to ensure that an evaluation takes place. Once the evaluation has taken place, you must swiftly convert the results into a plan of action and exert the leadership to develop and implement needed improvements. If you have done a good job, the evaluations will correctly identify needed improvements and provide you the opportunity to initiate corrective actions. The evaluation is the first of the steps to achieve plant profitability through improved maintenance performance.

Step 1—Develop a Policy for Evaluations

There must be a strong policy requiring that maintenance be evaluated on a regular, continuing basis. Such a policy will preclude the resistance that many poor maintenance departments exhibit to discourage evaluations. This policy will help to redirect the energy of resistance into an effort to prepare for evaluations instead. A typical policy might be:

> Maintenance makes a significant contribution to the overall profitability of our plant. To ensure that our services are effective we will evaluate maintenance on a regular, continuous basis. The evaluations will examine how effectively we perform our work, as well as how the plant supports and utilizes our services. Evaluation results will be publicized so that we may know where we stand, what we have done well and where we need to improve. I look forward to your continued cooperation.

Step 2—Provide Advance Notification of the Evaluation

Advise personnel about the evaluation and make a preliminary statement about its content, purpose, and use of the results. Eliminate surprises with the announcement but emphasize the policy of regular, continuing evaluations. The announcement should:

- Provide the dates of the evaluation.
- State the objectives.
- Remind personnel of the results of previous evaluations, commenting on accomplishments and further improvements needed.
- Explain how this evaluation is in line with your long-range policy to ensure continuous improvement.
- Thank everyone in advance and state that you are looking forward to their help.
- Offer to meet with anyone who may have problems with the evaluation content or dates.

Step 3—Educate Personnel on the Purpose of the Evaluation

Explain that the evaluation is a checklist describing what maintenance should be doing. Its results should describe how they did and provide a basis for developing an improvement plan. Promote the evaluation as an effort to identify what is done well—not what is done poorly. It is your opportunity to help your department account for its support of plant profitability objectives.

Emphasize the positive aspects of the evaluation in your educational effort. Change unfavorable misconceptions of evaluations by telling personnel they are the means of finding out how maintenance can do better. When preparing personnel for the evaluation, acknowledge that there may be a genuine fear of audits and evaluations. This fear may be based on previous bad experiences, or it may have no basis. Many maintenance workers simply don't want anyone looking over their shoulders. Other fears are fueled by managers who themselves may create uncertainty with remarks that mislead. One maintenance manager proclaimed that "Auditors were those who come in after the battle is lost and bayonet the survivors!" Such attitudes contribute little to the mental preparation that must be made if a plant is struggling to improve.

Some unions have been known to threaten strikes when an evaluation is announced. They fear that their "sacred promise" of preserving jobs will be threatened by possible work force reductions as the result of an evaluation. The fact that the evaluation can also lead to improvements that will make their work easier seldom occurs to them. It follows that maintenance supervisors—caught up in the threat of their job being made harder by the union position—may resist as well.

Avoid surprising personnel as this adds to resistance and creates distrust. People don't like surprises. Therefore, let personnel know what is coming and avoid resistance. The evaluation should be considered as a checklist describing what maintenance should be doing. Its results describe how they did and provide a basis for developing an improvement plan.

Resistance to being evaluated also occurs when there are several operations that will be evaluated. The concern is that performance between several maintenance operations will be unfairly compared when, in fact, they are not similar and would be better compared on another basis. Regardless, comparisons will be made. Therefore, be aware that the sponsor of dual evaluations must convey a supportive attitude. Don't be unduly concerned when you become aware that several plants are undergoing maintenance evaluations. Usually, you will find that the sponsor—usually a general manager—is eager to provide help in any areas that you lack resources (e.g., computer programming support). The general manager should provide encouragement to conduct the evaluation in the first place and should follow up to see that something constructive is done with the results. It remains only that you conduct the evaluation and ask for help.

In your own plant, management will want to know how well its policies are understood and how effectively the procedures based on those policies are being carried out. Let them know. Although they are concerned with the quality of the maintenance program they will be equally interested in learning how well, for example, production supports and cooperates with the program.

Therefore, you should be less concerned with what others may think. Get on with the evaluation and assume that every other party is interested, concerned and ready to help.

Step 4—Schedule the Maintenance Evaluation

Schedule the evaluation carefully—particularly if it will involve a physical audit lasting several weeks. In selecting the evaluation dates be aware of potential conflicts that might distort evaluation results, for example:

- If a shutdown has just been completed, there may be distracting start-up problems. Similarly, if a shutdown is coming up, preparation may compete for the attention of personnel.

- Peak vacation periods may find key personnel away from the plant. Be aware of the expected absences of key people and weigh their nonparticipation in deciding when the evaluation should be conducted.
- Recent personnel changes could limit knowledge of evaluation points and personnel cutback or staff reductions might affect attitudes.

In general, the evaluation should be carried out in a stabilized situation with as few distracting conditions as possible. With suitable advance notice, the plant can prepare for the evaluation and look forward to learning how it is doing.

Timing of evaluations is important. If they are conducted on a regular, continuing basis, people will look forward to them as an opportunity to demonstrate progress. If maintenance personnel feel that the evaluation is constructive, they will prepare for it without hesitation. In subsequent evaluations, if they accepted the evaluations and are convinced of their value, they will make a conscious effort to improve on previous results.

Step 5—Publicize the Content of the Maintenance Evaluation

Make sure you explain the content of the evaluation in advance. A maintenance evaluation covers such a variety of activities that it is unlikely any advance notice will constitute a dramatic shift in performance. By announcing what is to be evaluated, you will help organize the evaluation. Reports will be ready, personnel scheduled for interviews, and the evaluation can be carried out more effectively.

Agreement on evaluation content also clarifies expectations of both the maintenance department and the evaluators. For example, if productivity is to be measured, hourly personnel should be advised so they do not misunderstand the intent.

The scope of the evaluation includes three broad areas:

- The maintenance organization—Evaluate the following factors:
    - The efficiency of the maintenance organization in *controlling maintenance personnel* and carrying out its work.
    - The quality of *maintenance supervision* and its effectiveness.
    - The ability of maintenance to measure its *workload* and determine the correct workforce size and craft composition.
    - The degree to which maintenance effectively controls *labor utilization.*
    - The degree to which maintenance considers *productivity* important and acts to improve it.
    - The *level of motivation* of maintenance supervisors and workers.
    - The quality and effectiveness of *craft training.*
    - The quality and effectiveness of *supervisor training.*

- The maintenance program—Evaluate the following factors:
  - How effectively maintenance has *defined its program* so that everyone can act in a well-informed manner.
  - How well *maintenance terminology* has been defined, communicated, and used correctly.
  - The effectiveness of *preventive maintenance* in extending equipment life and avoiding premature failures.
  - The quality and effectiveness of the *work order system*—the communication link of maintenance.
  - Whether maintenance has the *necessary information* to identify work, control its performance and measure its effectiveness in reducing costs and minimizing downtime.
  - The quality of *maintenance planning* in assuring major jobs are carried out effectively with minimum resources and least elapsed downtime.
  - How well *performance standards* are used to ensure job quality and control labor use.
  - Whether *scheduling* assures the least interruption of production and the best use of maintenance labor.
  - The degree to which *work control procedures* assure effective labor, quality work, timely work completion, and knowledge of job status.
  - How effectively *maintenance engineering techniques* are used to ensure equipment reliability and maintainability.
  - The degree to which *technology* is employed by maintenance to help perform its tasks more effectively.
  - How well maintenance uses *backlog data* to determine whether they are keeping up with the generation of new work—and if the current workforce has the capacity to meet the workload.

- The environment in which maintenance operates—Evaluate the following factors:
  - Whether management has assigned a clear, realistic *mission* to the maintenance function.
  - The degree to which maintenance is supported by *policies* that ensure common understanding and compliance with most procedures.
  - How well management supports the *maintenance program* and causes it to be used effectively in the support of the production strategy.
  - How well *production* understands the maintenance program, cooperates and uses its services effectively.
  - Whether *staff departments* such as purchasing understand the maintenance function and support it with a high level of service.

- The effectiveness of *material control* in support of the maintenance function.
- The degree to which *safety procedures* are followed and unnecessary accidents or injuries avoided.
- How well maintenance carries out *housekeeping tasks* to help preserve equipment and create a neat, clean working environment.

The environment is the attitude of the plant toward maintenance. When poor, it is the most common reason for resistance to evaluations. Even a first-rate maintenance organization with an effective program can fail if the environment does not provide receptivity and support. Examining the environment includes answers to some difficult questions regarding:

- the quality of management sponsorship
- the degree of operations cooperation
- the level of service provided by staff departments such as purchasing or stores

Not surprisingly, when maintenance personnel realize that you are also evaluating key factors over which they have little control—but which influence their performance—their resistance diminishes.

Be aware that an evaluation of the plant environment can reveal an uncooperative operations department or an unsupportive purchasing agent. Therefore, it will be necessary to ask the support of your plant manager in conditioning these other departments for their participation as well as helping to improve the environment.

Examination of the environment can include some steps that your plant management must take to ensure the success of the program. Typically, these steps include a clear mission and realistic policies to guide other departments in how they must support your program. Although you can only advise your plant manager of these needs, you will invariably find that he or she appreciates your concern and will act positively to meet your department's needs.

Step 6—Use the Most Appropriate Evaluation Technique

Evaluation techniques should be considered based on the plant situation. Some operations may require an evaluation in which every detail must be scrutinized; other operations, having established the essential pattern of evaluations, may simply check progress by measuring only a few critical areas.

There are three techniques that you can use to evaluate the performance of your maintenance department:

- A physical audit
- The questionnaire
- The combined audit and questionnaire.

Physical Audit

A physical audit is usually conducted by a team. Generally, the team is made up of consultants or company personnel (or both). They examine the maintenance organization and its program as well as activities that affect maintenance (such as the quality of material support). The physical audit must examine the total maintenance activity at first hand, by observing work, examining key activities (such as preventive maintenance, planning, or scheduling) reviewing costs, and even measuring productivity. It relies on interviews, direct observation of activities and examination of procedures, records, and costs.

When done properly, the physical audit produces effective, objective and reliable information on the status of maintenance. The physical audit by itself should be used when you feel that personnel could not be frank, objective, and constructive in completing a questionnaire. If the physical audit technique is used, be prepared to spend several weeks conducting the evaluation. The evaluation can be disruptive because of the time required to help explain procedures or participate in interviews. Therefore, it must be well organized in advance.

Questionnaire

A questionnaire gives a cross-section of randomly selected plant personnel an opportunity to compare their plant's maintenance performance against specific standards. This cross-section might include personnel from management, plant staff departments (such as purchasing), production, and maintenance. Appendix A (at the end of this chapter) illustrates the type of standards against which preventive maintenance, for example, should be evaluated.

Although a questionnaire is subjective, the results are, nevertheless, an expression of the views of the plant personnel. Therefore, participants have committed themselves to identifying improvements they see as necessary. To most of them, this constitutes potential support for the improvement effort that must follow.

When administering a questionnaire, be careful to ensure that personnel are qualified to respond. For example, participants outside of maintenance must respond only to those standards on which they have personal knowledge. The evaluation must be administered so that questions on evaluation points can be answered completely.

When carried out properly, the questionnaire can produce reliable results quickly while minimizing disruption to the operation. The questionnaire has the advantage of being administered often so that progress against a "benchmark" can be measured. For example, one plant was able to establish areas in which improvement was still needed while setting aside areas in which good progress had been attained as the result of a continuous series of evaluations.

The questionnaire is the best choice when a quick, nondisruptive evaluation can serve as a reasonable guide in developing an improvement plan. It must be carefully crafted so that it embraces all of the elements of the maintenance program—like PM—as well as those activities that affect the program—such as purchasing. A cross-section of 15 percent of the plant population—including production personnel (your customers) and staff personnel (like accounting)—is adequate to produce good results.

Participants should have personal knowledge of maintenance performance for the standards against which they are comparing maintenance. There should be selectivity in who responds to what. For example, the accounting manager could evaluate the quality of labor data reporting but could not evaluate housekeeping. Questionnaires are rarely of value if they are not administered in a controlled environment. Typically, if distributed for completion at the respondents' leisure, expect poor results because there are too many opportunities for misunderstanding. Questionnaires administered in a controlled environment where participants can be oriented and their questions answered is always best.

Physical Audit/Questionnaire

The combination of a physical audit and a questionnaire provides the most complete coverage. The techniques work together—for example, the questionnaire provides confirmation of physical audit findings. This technique is preferred by consultants and corporate teams because it combines the objectivity of the outsider in the physical audit, while the questionnaire helps to educate personnel and gain their potential early commitment to improvement. As the insiders, their help is needed. Without it, little will happen. This combined technique is the best way of preparing for the improvement effort that must follow.

Step 7—Announce the Results of the Maintenance Evaluation

By publicizing the evaluation results, there will be clear evidence that you acknowledge both the good and the bad. More importantly however, you also exhibit a commitment to do something. By sharing the results of the evaluation with your personnel, you confirm that you expect their help in attaining improvements. Never keep the results a secret as this will decrease credibility and make improvement actions more difficult. Evaluation results should be discussed openly and constructively so that the personnel who must later support improvements are being brought along.

Don't rationalize that you are solely responsible when the evaluation reveals a poor performance. You probably got a lot of help from many people in the resulting poor performance. For example, your plant management may not have provided the necessary policy guidelines to ensure the maintenance program can be carried out in a positive environment. Or production may have required so many equipment changes that you were left with inadequate staff to maintain existing equipment.

Similarly, you should not blame your supervisors for disappointing results. Little is gained by trying to fix blame. Maintenance reaches into so many areas that few people would be without some degree of responsibility for the decline of maintenance performance. Always look ahead—determine current performance level and move forward from there.

Remember, maintenance is a service; you cannot compel operations to follow the program. You can only offer the service and hope for cooperation and support. Work toward creating an environment in which you can win support and co-operation. By including operations and staff departments in the evaluation, you not only examine aspects over which you have little control, but you identify what they can do to help you.

Plants performing poorly often do not show results to the personnel who participated. Better plants not only shared the results but sought help in interpreting the results and soliciting recommendations. For example, one successful maintenance manager observed that, "whatever the current performance is, it didn't get that way overnight." He approached his task of improving maintenance performance by saying, "since this is what we think of our maintenance program then let us now consider what we must do about it." This plant was already on its way to improving because the support and enthusiasm for doing better had been successfully harnessed even before the evaluation was completed. There was involvement and it showed.

Most plant management personnel are uncertain of the roles they must play in creating an environment for a successful maintenance program. They acknowledge that stating a clear mission, providing policy guidelines, and requiring a definition of the maintenance program were useful approaches to the creation of a positive environment for maintenance.

Production personnel generally did not realize the comprehensive nature of the maintenance program and how much its success depended on their support and cooperation. Staff in purchasing, warehousing, accounting and MIS express a need to better understand the maintenance program so they can support it more effectively.

Many maintenance managers acknowledge that their own maintenance programs are not adequately defined much less properly explained to either maintenance personnel or the rest of the plant. Some admit a need to align basic program elements like PM, planning, scheduling, and maintenance engineering. Evaluations often reveal facts about the maintenance program that were concealed by defensive information reaching managers or the misleading attitudes taken by maintenance personnel.

Step 8—Take Immediate Action on the Results

The most convincing way to demonstrate that the evaluation was a constructive step is to organize an improvement effort immediately. You must commit to a constructive use of the results by converting them into an improvement plan and immediately organizing the improvement effort. This is the main objective of the evaluation. If the evaluation is one of a series, results should be compared with the previous evaluation. This demonstrates progress as well as the identification of areas that need more work. Separate the good from the bad. Offer congratulations on the good performances and organize the activities requiring improvement into priorities. Actively solicit help from anyone capable of providing it. Most will participate willingly.

Thereafter, you should develop a plan for further improvement and implement corrective actions. If there are corrective actions beyond your capability, don't hesitate to seek help. Plant managers are usually pleased to be asked to help. It is also gratifying to learn that corporate managers, particularly those responsible for multiplant operation, are eager to help as well. Set up an advisory group and get underway. Let them first determine why certain ratings were poor. Then, ask for recommendations for improvement. Change the members of the advisory group frequently to encourage different views. As recommendations are made, try them in test areas before attempting plant-wide implementation.

Step 9—Announce Specific Gains Resulting from the Maintenance Evaluation

As soon as any gains that can be attributed to the evaluation can be identified, you should announce them and give credit to the appropriate personnel. People like to know how they did. Tell them. In the process, your candor will invariably encourage a greater effort in future evaluations.

Step 10—Specify the Dates of the Next Maintenance Evaluation

Specify the dates of the next evaluation. As necessary, identify any additional activities that will be evaluated. Establish new, higher performance targets for the next evaluation. Reinforce the policy of continuing evaluations.

HOW TO USE EVALUATION RESULTS TO IMPROVE MAINTENANCE

There is a valuable lesson in observing how maintenance organizations have successfully converted evaluation results into improved performance. Case studies describe evaluation results and their conversion into corrective actions.

Improving Supervisor Training

One of the areas that evaluations most frequently find in need of improvement is supervisor training. Most plant personnel described a nonexistent supervisory development program. There were major problems:

- Newly appointed supervisors were expected to figure out their duties for themselves. Training on the maintenance program was often vague because the program itself was seldom well-defined and documented.

- Supervisors identified for promotion were seldom trained on the broader duties they would be required to perform.
- There was little time allocated for supervisory training of any kind.

In the little training given, emphasis was given to technical skills, rather than "foremanship." As a result, many supervisors were reluctant to move from "one craft" crews to the "multicraft crews" necessary for more flexible organizational arrangements like area maintenance.

There was an impression that management saw supervisors as those who "fixed equipment" rather than those who managed the efforts of craftsmen in carrying out the maintenance program.

Generally, supervisors were criticized for their lack of supervisory skill. Few had ever received any supervisory training. They were rated well on the technical competence which they carried into their supervisory jobs as they moved from craftsmen to supervisor. Most supervisors displayed a tendency toward working directly with their crews rather than leading them. Generally, superintendents were criticized for acting like foremen rather than managers. In turn, supervisors were criticized for acting like workers. Supervisors were often labelled as "tire-kickers" rather than managers. Poor marks were given for control of work and utilization of labor.

Solution:

- Establish a policy for supervisor training that emphasizes foremanship.
- Verify that there are no impediments precluding attendance at training such as an inadequate number of supervisors.
- Require completion of the prescribed curriculum and then assess supervisory performance.
- Emphasize the training of new supervisors.
- If in-house training is not available, arrange attendance at suitable commercial training.
- Get involved in setting up the training, monitoring it, and checking progress.
- Do not accept excuses for nonparticipation in the program.
- Reconsider supervisor selection criteria and look for talent among those without maintenance backgrounds.
- Require emphasis on foremanship in supervisor training.
- Demand that supervisors utilize crew members more effectively on repairs and assume more responsibility for work control.
- Initiate a supervisory selection criteria so that those with limited management talent will not stagnate the line of progression with people who are "dead-ended" when they reach the supervisor level.

Monitoring and Improving Craft Training

By contrast with supervisor training, craft training programs were rated well and considered successful. Contractual requirements that training for hourly craft personnel be provided was often credited with their greater success. However, some craft training tended to emphasize what craftsmen felt they needed rather than what the maintenance situation required. One plant learned that crafts that would never use infra-red testing techniques were receiving training on it instead of what their program required.

Solution:

- Require regular reports on craft training progress and periodically review its content to ensure it remains consistent with needs.
- Attend a sampling of the training to demonstrate personal interest and monitor its content.
- Talk with craftsmen to determine their level of satisfaction with the training.
- Periodically, have a training needs analysis conducted to ensure continuing adequacy of the program.

Defining the Maintenance Program

Few maintenance departments had a solid, well-publicized definition of the way they carry out their program and what they expect of their own personnel as well as their customers. Program definition should explain how maintenance services are requested, organized, executed, controlled, measured, and so on. There was confusion on how to do these basic things. Often, operating personnel acknowledged no role other than to submit work requests. There was little joint scheduling of major jobs and confusion on what was to be done and when. Performance on major jobs was seldom questioned by operations nor were overall maintenance costs. Results usually indicated that since maintenance did not advise anyone how they operated, the misunderstanding was expected.

Solution:

- Check to see if the maintenance program is defined.

- Verify that the definition has been explained to maintenance personnel as well as their customers.

- If there is no definition, develop one, publish it, and offer to educate plant personnel on it.

- Follow up to verify that the program definition is understood and followed.

Developing a Maintenance Mission Statement

Many maintenance departments had vague missions like "support operations" or "get the product out." Those examined were compared with a reasonable standard such as:

The primary objective of maintenance is to maintain equipment, as designed, in a safe, effective operating condition to ensure that production targets are met economically and on time. Maintenance will also support nonmaintenance project work (like construction) as the maintenance workload permits. In addition, maintenance will maintain buildings and facilities and provide support services such as hoist-operation or power-generation.

Generally, the absence of a clear maintenance mission statement confused the maintenance program. Typically, maintenance, instead of focusing on preserving equipment, ensuring quality product and dependable equipment was, too often, diverted into nonmaintenance activities like: performing "process optimization" (usually improperly engineered modifications thought to make the process better) or moving, upgrading, modifying and installing equipment—or construction.

There was confusion in differentiating between maintenance and nonmaintenance work not only within maintenance but plant-wide.

Maintenance is the repair and upkeep of existing equipment, buildings, and facilities to keep them in a safe, effective, as-designed, condition so they can meet their intended purpose.

Maintenance is an operating expense. Nonmaintenance which includes actions such as construction, equipment installation, modification, or relocation is generally capitalized, depending on the cost.

Maintenance departments without clear boundary lines and specific policies regarding the division of maintenance and nonmaintenance work usually failed to carry out the basic maintenance program adequately. Some maintenance managers with responsibility for maintenance helped create the problem by giving undue emphasis to nonmaintenance project work.

Results further revealed that some plants had what appeared to be low-cost maintenance programs. However, in reality, maintenance resources were misused on projects, often over the objection of the maintenance. Typically, deterioration of equipment resulted from a lack of maintenance and it required premature replacement. An unfavorable maintenance budget was due not to excessive maintenance cost but rather, to nonmaintenance work that was expensed rather than capitalized. Maintenance was often the unfortunate "scapegoat."

Solution:

- Review the current maintenance mission to ensure it clearly establishes the conduct of the maintenance program as the department's primary responsibility.
- If maintenance is required to perform nonmaintenance work, ensure there are safeguards to preclude improper utilization of maintenance labor.
- Make sure the maintenance supervisors are not victimized by unrealistic emphasis of nonmaintenance work, such as modification.
- Verify that maintenance supervisors are not "construction types" who would rather get involved in nonmaintenance activities like construction or modifi-

cation while neglecting their primary task of maintenance—the repair and upkeep of existing equipment.

Establishing Useful Maintenance Policies

Maintenance, as a service, cannot compel production to comply with its program. Consider therefore, an ambitious operating superintendent, who in his or her quest for quality, high output, and so on, forces maintenance personnel into "process optimization" rather than allowing them to carry out the basic maintenance program. Certain "process optimization" actions (usually modifications) may not be feasible or even necessary. Many should be approved by engineering but aren't. Some that should be capitalized may be chopped up into incremental expensed actions to avoid the "gauntlet" of getting capital funding approved. Should maintenance object, they are reminded that their indirect service role does not make a profit. Using this kind of leverage, the ambitious operating superintendent may get his way—right or wrong. In the process, he may, without malicious intent, diminish the maintenance program.

Abuses of the maintenance program also include ignoring published maintenance schedules in the interest of meeting production targets. Often, this noncooperative attitude permeates the operations department. Operators abuse equipment. Shift supervisors act innocent, knowing units require servicing or repair but hope the next shift will have to cope with the breakdown. There appears to be unfair charging of the resulting downtime to maintenance. These scenarios are prevalent and they undermine the maintenance program.

Evaluations should ask directly whether there are policies which could prevent this from happening. The results will probably show that such policies, which can only be issued by plant management, are weak or missing. As a result, maintenance may feel that operations views the maintenance program less seriously than they should. Uniformly, very few managers clarified production responsibilities for utilizing maintenance services effectively.

Solution:

- Seek plant management help in obtaining policies to preclude these abuses of the maintenance program.
- Ask plant management to consider making operations responsible for the cost of maintenance. They will become more demanding of quality work, completed on time and kept under budget. Expect a dramatic reduction in downtime.

Standardizing Maintenance Terminology

Few maintenance departments adequately define terms they use every day. Some vital terminology is often misleading even within maintenance. The workload for maintenance often cannot be identified, much less measured. Especially confusing may be the meaning of preventive maintenance, emergency repairs and planned, scheduled maintenance. Only a few workers may be able to distinguish the backlog from the open work order file. Modification versus corrective maintenance and overhauls versus rebuilds may be confused. In one plant, there were seven different versions of the meaning of PM just within maintenance.

The absence of adequate definition of terms often emerges in other areas such as PM, planning and so on, where proper definitions are necessary to spell out effective procedures. Outside of maintenance there may be confusion on basic terms.

Solution:

Check the existence of a basic terminology. If it does not exist, demand it. If it does exist, make sure it has been published and verify its correct use. Typical of the type of terminology that must be defined is the basic work that maintenance performs. Appendix B at the end of this chapter provides a glossary of basic maintenance terminology.

How to Establish the Maintenance Workload

Identification of the essential work to be performed by maintenance and the determination of a work force of the proper size and craft composition are often poorly carried out. One maintenance superintendent determined work force size by adding staff until the overtime went down—hardly adequate but typical of the lack of regard for this matter. Most maintenance departments do not know how to go about this task nor do they have elements in their information systems which permit them to confirm or adjust current workforce levels.

Solution:

Require maintenance supervisors to verify the size and craft composition of their work forces regularly. If they are hesitant, chances are good they have no idea how to do it. It is often necessary to demand they do it, then guide them through some reasonable procedure. For example, using the definition of the workload as a starting point, estimate the amount of labor for PM and planned and scheduled maintenance tasks. Then, make reasonable allowances for emergency repairs and unscheduled work. Next, identify the number of labor hours for routine activities such as building repair. Assemble the results, comparing them with typical industry standards:

Preventive Maintenance	10%
Planned and Scheduled Maintenance	50%
Unscheduled Repairs	20%
Emergency Repairs	10%
Routine Activities	10%

Once these levels have been established with estimates, utilize the information system to confirm how labor is actually used and make corrections in the distribution. Then adjust the workforce size and craft composition.

Measuring the Productivity of Maintenance Workers

Worker productivity is seldom measured in some organizations, nor are steps taken to establish its importance. Hourly personnel may be fearful of productivity measurements. Most plants acknowledge that productivity was low, had not been measured and that the quality of labor control would not yield necessary improvements. Invariably, the maintenance supervisor could have improved productivity by giving better work assignments, ensuring work was preorganized and spending more time supervising his crew. Why he did not was seldom investigated.

Solution:

Make productivity measurements mandatory. The most effective techniques of random sampling, in which the pattern of what personnel are doing is very informative. For example, calculate what percentage of time is spent:

- Working
- Idle
- Travelling
- Waiting
- Performing clerical functions

Also critical is when personnel are doing these things. For example, if personnel are idle at the start of the shift, it usually means that work assignment procedures are inadequate. Similarly, if they are performing clerical tasks at the end of the shift, it often means that labor reporting techniques should be revised.

Productivity measurements are the most effective way of verifying the quality of labor control. However, it will be necessary to sit on most maintenance departments to make this happen. Yet, it must be done. The blow can be softened by allowing maintenance to conduct self-evaluations.

> *Case Study:* One maintenance department made major strides in improving productivity by measuring the factors that inhibit work. The results indicated the degree to which factors such as the difficulty of obtaining materials robbed personnel of productive time. These results provided very specific improvement targets that responded to correction easily.

Beware of maintenance supervisors who consider anyone breathing to be 100 percent productive. Educate them. Find out what these supervisors are doing and how well they actually control their crews. Their supervision time is critical to productivity, and they must spend at least 60 percent of their time on active supervision to assure at least 40 percent productivity (not very good). Therefore, make sure your supervisors have enough time to supervise. If you have heaped administrative tasks on them, the poor productivity of their crews may be your fault.

Conducting an Effective Preventive Maintenance Program

The biggest problem with Preventive Maintenance (PM) is a misunderstanding of what it is. Views range from "everything you did before equipment failed" to "the whole maintenance program." This confusion can carry over into the effectiveness with which PM is scheduled, executed and benefits gained. Invariably, those who define PM as routine, repetitive actions (i.e., inspection, lubrication, testing) have the best control over its effectiveness. Those who do poorly in PM also do poorly in defining terminology.

Many PM programs can not be administered effectively because inspection and testing are mixed with repairs. Thus, the deficiencies resulting from inspection and testing which should be separated and classified into emergency or unscheduled repairs and planned maintenance are not. As a result, the identification of workload elements is blurred and precluded determination of the proper workforce size and craft composition.

Generally, good use is made of nondestructive testing (like vibration analysis). However, such efforts are not always integrated into the overall PM program.

Solution:

Require that maintenance personnel present their preventive maintenance program to operations and management. If they can't, they don't have one. If they hesitate, they need to have it challenged or evaluated. Most likely, they may not know what to do or how to organize PM.

> *Tip:* PM services, like inspections for fixed equipment, must be organized into routes in which equipment in each is checked at fixed intervals. Inspections should be detection oriented and you should verify that they are completed. After inspection, deficiencies must be reported and decisions made on their importance before they are converted into new jobs.

Although this organizational concept is simple, it is often overlooked. PM is too important to ignore. Your personal attention is essential.

Establishing More Effective Maintenance Planning

Planning is generally not a well-executed function. Among the problems:

• There are no criteria describing what work should be planned.
• Planning procedures are poorly described.
• Information for control of planned work is sparse.
• There are not enough planners.

- Planners are often misused on unscheduled or emergency work or as relief supervisors. Few are adequately trained.

Many maintenance departments consider work to be planned if there is a pause between work request and work execution. Most play with statistics to arrive at a plausible percent of work they claim to be planned. Usually, there is a faulty classification procedure in which it is difficult for anyone to determine what work is planned. Often, the routine, repetitive PM services (such as weekly inspections or monthly lubrication tours) are considered planned when, in fact, they are merely scheduled. Many large planning departments have failed to cut back their staffs when the use of the computer displaced considerable administrative work formerly done by planners.

Solution:

- Examine the planning procedure.
- Check to see if there is a criteria for determining which jobs will be planned and scheduled.

Establishing a Responsive Work Order System

Some organizations do not use a work order system. Usually, there is one piece of paper—inadequate for detailed planning but overly complex for a simple request. No provision is made for the inevitable verbal orders. Efforts to suppress them result in no work order and no information about the job. Standing work orders become a burial ground for costs and repair history that should have been isolated unit-by-unit. Rarely is there an engineering work order to control non-maintenance project work (like construction). Personnel may try to use the maintenance work order to control this work when performed by a contractor.

Solution:

Organize a taskforce made up of accounting, operations, material control, and maintenance personnel to evaluate and correct the inadequacies that often emerge under scrutiny. Typical shortcomings:

- There is no work order system. Rather, maintenance is attempting to use one piece of paper for all types of work. The result is that big jobs are poorly controlled and simple jobs are made a big administrative deal—then ignored.
- There is an inadequate linkage with accounting data resulting in no information on costs.

Maintenance will welcome the evaluation because the work order system is an extension of the accounting system and their essential communication link. It is one of those items that affect maintenance performance over which they have no control.

Ensuring Accurate Field Labor Data

Labor use reported by hourly personnel is often carelessly recorded with more attention paid to "filling up the 8 hours" than providing accurate, timely, and complete data. Some supervisors resort to filling out craftsmen's time cards themselves to get "something better than what they were receiving." Most supervisors admit that the labor data they record is largely conjecture when they have 10-12 people in the crew each performing 6-8 jobs per shift. Hourly personnel readily admit they are the best source of information on what they actually did. They should be given opportunity to report it directly. Often crew members do not report accurately because:

- the importance of reporting is not explained;
- they are suspicious of the reports use.

The end result is missing or misleading information because there is poor field labor data to support it. Most maintenance departments do poorly in reporting labor because of a lack of emphasis rather than a faulty reporting scheme.

Solution:

- Ensure labor is reported initially by those who do the work and not by the supervisor who guesses at it. The supervisor should verify, not initiate, the information.
- Verify the accuracy, timeliness, and completeness of labor data by comparing it with actual work assignments.
- Solicit the accounting viewpoint in finding ways to streamline labor reporting.

Remember that the only way maintenance can reduce the cost of doing its work is to improve its effectiveness in installing materials. This makes the control of labor a vital function and one that requires careful scrutiny.

Getting the Best Material Control Support

Material Control (i.e., inventory control and direct charge purchases) is usually well-carried out when managed by material control professionals. When maintenance controls parts inventory, however, there is little continuity between inventory control and direct charge purchasing. Usually, it splits the material control function and creates confusion. Poorly administered maintenance programs usually produce poorly administered inventory-control programs when maintenance is responsible for the function.

A major problem is the adequate identification of parts. Often supervisors are forced into this role because hourly personnel feel procedures were difficult or time-consuming. As a result, parts identification tasks seriously reduce the time supervisors should have been in the field controlling work. In turn, this situation produces lower worker productivity.

Solution:

- Check maintenance downtime against causes such as no materials available, wrong materials issued, and waiting for materials. These factors are the best indicators of the state of material control.
- Make sure the material control function is not split—with purchasing under accounting and stores under maintenance, for example. It seldom works.
- Watch a few stock room transactions to learn how well maintenance craftsmen actually are able to identify materials.
- Check the supervisor's office. If it looks like a library of parts books, this clerical activity is probably his or her principal job.
- Check the "bootleg" storage areas if you wish to learn the awful truth of why material costs are so high.
- Check the foreman's desk drawers and wall lockers if you want to find out where the real stockroom annexes are.
- Don't threaten a forced clean up, as too many usable items will end up in the waste dump rather than returned to the storeroom.

The resolution begins with a physical check such as the one outlined above. However, the solution lies in a serious joint corrective effort between maintenance, accounting, purchasing, and the stockroom staff. Within maintenance, instruct everyone on how the material control program works. Often they don't understand it.

Using Maintenance Information Systems Effectively

Information systems seldom cover decision-making information aspects adequately. This information includes: cost, repair history, backlog, labor utilization, the status of major jobs and performance indices such as maintenance is often cost per unit of product. There is an often an overabundance of administrative information on items such as parts cross-references, equipment lists or personnel who were absent. Thus, while everyone can easily look up parts, few can adequately manage maintenance because information systems emphasize administrative rather than management information.

Systems provided by the corporate level, in an effort to achieve uniformity between several plants, do not work well. Some maintenance organizations purchase standalone software packages only to find that they need an integrated system but cannot obtain local support to create necessary communications software. Some package programs are incompatible due to both language and logic considerations. Often commercial packages fail to link adequately with field labor and material data from time cards and stock issues or purchasing documents. Most organizations are not satisfied with the timeliness, completeness and accuracy of information.

Solution:

Specify the performance indices required by management such as cost per unit of product. These will equate to specific information required to create the indices. Make certain this information exists. Ignore the clerical and administrative information such as parts identification and equipment specification. These are merely the means by which maintenance carries out its internal communications. Virtually every package program has these in great profusion.

Instead, concentrate on whether maintenance actually has decision-making information to manage the overall function. Mandatory are:

- labor utilization—how effectively each supervisor controls the crew and whether absenteeism is properly controlled.
- backlog—the degree to which maintenance keeps up with the generation of new work.
- costs of units and components—cost data by which troublesome equipment can be identified.
- status of major jobs—for selected major jobs, be able to determine cost and performance.
- repair history—the chronic, repetitive problems and failure trends that must be found and corrected.

Once the right information has been obtained, ensure it reaches the people who need it to make their decisions. If required, instruct them how to use it.

Optimizing Maintenance Engineering

Maintenance engineers are usually used on nonmaintenance engineering projects like equipment installation or modification. Not enough attention is paid to functions that ensure equipment maintainability and reliability. A large percent of maintenance engineers are young and have degrees in a specific engineering discipline. They tend to gravitate toward what they understand best (project work) rather than maintenance engineering. Plants contribute to this problem by failing to adequately define maintenance engineering.

Solution:

- Require that maintenance engineering focus on ensuring the reliability and maintainability of equipment.
- Ensure that practical solutions like effective PM or repair standards are being instituted to reinforce reliability objectives.
- Verify that a procedure exists to determine that newly installed equipment

is able to be maintained. Does the equipment have maintenance instructions, spare parts lists, and wiring diagrams, for instance.

- Assure that maintenance engineers are not acting as project engineers or misused on nonmaintenance activities such as construction.

Supporting Engineering Projects with Maintenance Resources

Invariably, major engineering projects are carried out with maintenance personnel rather than by a contractor or a segment of maintenance maintenance workforce set aside for this work. Too often, this work is not controlled and the maintenance program suffers from a shortage of available labor. Usually these projects are well planned and organized because they are in the plant spotlight. At the lower end of the cost scale, proper funding is questionable with frequent attempts to expense the work against the maintenance budget rather than seek capital funds through proper channels. Projects tend to be executed on time but often at the cost of diverting maintenance labor from necessary maintenance work. Most projects have direct attention from management—and monitoring by corporate personnel. These aspects, in part, explain their success.

Solution:

- Curb the tendency to misuse maintenance resources on nonmaintenance projects by developing and enforcing policies that preclude the unfair diversion of maintenance labor from its own program.
- Check lower cost project funding practices to make sure maintenance is not unfairly paying the tab for a project that should have been capitalized.

Ensuring Support for Maintenance by Operations, Staff Departments and Plant Management

These departments show an appreciation for the value of a good maintenance program but, lack a specific means of providing support beyond saying "I'm for good maintenance."

Solution:

- Document specific instances that operations have caused problems by not making equipment available when maintenance is due to be performed.
- Visit with the production superintendent and suggest how the situation may be improved through positive solutions such as a weekly, joint scheduling meeting.
- Measure how well staff services such as the warehouse are supporting you and, if you are dissatisfied, go to them and suggest how they might improve.
- If you find that management needs to step in, suggest policies that could help classify the supporting role of other departments.

Attaining Better Housekeeping

There is a tendency in maintenance to allow discarded components to build up, creating junk piles. There is poor follow-up in getting components into the rebuild pipeline—especially if done commercially.

Solution:

- Extend housekeeping inspections into the well-known areas where maintenance has a tendency to collect junk in case they might be able to use it later.
- Make sure the procedure for rebuilding components is well organized. Worn parts should be tagged before they are sent to the warehouse for classification. Once returned from rebuild, be sure parts go back into the warehouse. When installed, track their performance to ensure the quality of rebuilds.

Motivating Maintenance Personnel

Hourly maintenance workers usually requires strong motivation. However, motivation at the supervisor and planner level is usually marginal. Maintenance managers and superintendents may appear well motivated.

Solution:

- Increase direct contact with supervisors and workers to uncover the reason for their poor motivation. You will create more opportunity to discuss common problems like working conditions and safety practices. Often, solutions will surface and morale will improve.
- Be aware that motivation may be affected by administrative procedures such as the ease of parts identification. Check these matters and correct them.

HOW EVALUATIONS CAN IMPROVE PLANT PROFITABILITY

In this era of intense competition, only profitable plants will survive. Maintenance, which often represents over 30 percent of operating costs, is one of the few major costs that can be controlled. If costs are not controlled and reduced, they deny profitability directly by costing too much. An ineffective maintenance program creates yet another unfavorable cost impact: equipment downtime. Downtime is 2-1/2 times as costly as the maintenance that could have avoided it. The lost product never reaches the marketplace and the downtime from poor maintenance diminishes profitability.

An evaluation establishes the current maintenance performance level by identifying those activities needing improvement as well as those being performed well. The evaluation not only confirms performance but, is the starting point for any improvement effort. Yet, the most important by-products of a well-conceived and effectively conducted evaluation are the education of plant personnel and their commitment to provide genuine support for improving maintenance. An evaluation that successfully points the way to improvement makes that improvement more attainable and is a matter of pride to the maintenance. The very people who can create the improvements to assure profitability are ready to help in a cause they believe in.

Appendix A

Sample Questionnaire for Evaluating the Maintenance Function

The excerpt shown below illustrates a typical section of a self evaluation in which a cross section of plant personnel rate maintenance performance against a series of performance standards. The complete evaluation covers 33 areas affecting maintenance performance and includes 425 performance standards, but only preventive maintenance is illustrated here. Participants have a structured answer sheet which precludes their answering unless they have personal knowledge. Each standard is rated on a scale of 1 to 10 (highest) or X (I don't know). The evaluation requires about three hours, and responses are scored overnight by computer to provide opportunity to discuss results the next day with participants. Ratings identify not only the degree of improvement needed but the X (I don't know) scores spell out where more training or better communication is needed.

Preventive Maintenance (PM)—The PM program should successfully extend equipment life and avoid premature failures through timely inspection, testing, lubrication, cleaning, adjustment, and minor component replacements. As a result, there should be fewer emergency jobs and more work should be planned. As the planned work is performed, maintenance personnel will work more productively and the results will have lasting quality.

On a scale of 1 to 10 (10 being highest), rate the organization, execution, and effectiveness of preventive maintenance according to the following criteria:

- There is an effective overall PM program.
- Plant management understands and strongly supports PM.
- The PM program is oriented toward uncovering deficiencies before equipment fails.
- The PM program emphasizes safety.
- The PM program emphasizes the preservation of assets.
- There is evidence that the PM program has reduced the amount of emergency work.
- As the result of the PM program, more work is being planned.
- The staff required for each PM service and for the entire PM program is known.

■ PM services are verified for quality and adherence to the schedule.

■ New equipment is added to the PM program promptly.

■ The PM program is reviewed periodically and updated to reflect changing conditions.

■ Maintenance personnel conduct PM services effectively.

■ Maintenance supervisors ensure PM services are performed effectively and on time.

■ The PM program has been explained to operating personnel to enable them to cooperate and use its services effectively.

■ Operating personnel cooperate with the PM program and perform simple, routine PM-related tasks to help ensure dependable operation of equipment.

■ Appropriate nondestructive testing techniques (such as vibration-analysis and infra-red testing) have been identified and integrated, as required, into the PM program.

■ Each PM service has a standardized checklist which explains how and when the service is to be performed.

■ Each PM action is identified by a code or number to aid in scheduling, control and reporting.

■ Extensive repair actions during the conduct of PM services, especially inspections, are avoided.

■ The timing of PM services is carefully regulated according to fixed time intervals (fixed equipment) or accumulated operating hours, miles, and so on, (mobile equipment).

■ PM services for individual units of fixed equipment are linked together in routes to avoid unnecessary travel time or back-tracking.

■ PM services for mobile equipment are scheduled in advance to avoid unnecessary interruption of operations.

■ Individual operators and maintenance workers cooperate in the conduct of PM services.
Typical of the scoring provided by such an evaluation are the scores shown below for one of the standards of the preventive maintenance area.

■ The PM program has been explained to operating personnel to enable them to cooperate and use its services effectively.

Scores are provided for each of the standards in the following format:

	Management	Production	Maintenance	TOTAL
R	65	43	61	56
X	09	00	21	15

The reader would interpret the scores to show that management rated this standard higher than either production or maintenance, but 9 percent of the management group did not know. On the other hand, production—to whom the program was to have been explained—rated it much lower than maintenance and all of them knew that the explanation was not adequately provided (X = 00). Maintenance (R = 61%), thought they provide a good explanation, but 21 percent didn't know. The scores show that:

- Operations personnel needed a better explanation of the PM program.
- Management was not aware whether the explanation had been provided.
- Maintenance personnel felt they did a better job than operations did—but 21 percent did not know.

　　Percentile scores are also awarded for each of the 33 areas and the overall evaluation.

Appendix B

Maintenance Terminology
(Workload Definition)

Maintenance work consists of the repair and upkeep of existing equipment, facilities, buildings or areas in accordance with current design specifications to keep them in a safe, effective condition while performing their intended purposes. Categories include:

Preventive Maintenance (PM)—Equipment inspection and non-destructive testing to determine future repair needs and their urgency. Lubrication and minor adjustment to prolong equipment life. Cleaning, adjustment, and minor component replacement to help ensure dependable operation.

Unscheduled Repairs (Running Repairs)—Unscheduled, nonemergency work of a short duration. Work that should be accomplished within approximately one week with little danger of equipment failure in the interim. Such repairs are usually performed by one craftsman, often in two hours or less.

Emergency Repairs—Immediate repairs needed as a result of failure or stoppage of critical equipment during a scheduled operating period. Imminent danger to personnel and extensive equipment damage, as well as substantial production loss, will result if equipment is not repaired immediately. Scheduled work must be interrupted and overtime, if needed, would be authorized in order to perform emergency repairs.

Scheduled Maintenance—Extensive major repairs such as rebuilds, overhauls or major component change-outs requiring advance planning, lead time to assemble materials, scheduling of equipment shutdown to ensure availability of maintenance resources (i.e., labor, materials, tools, and repair facility space) and management of the job from inception to completion. Most of this work would be accomplished during complete or partial shutdowns.

Routine Maintenance Activities—Shop clean up, training, safety meetings, tool repair, buildings and grounds work.

Non-maintenance Work consists of:

Construction—The creation of a new facility or the changing of the configuration or capacity of a building, facility or utility.

Equipment Installation—The installation of new or rebuilt equipment.

Equipment Relocation—Repositioning major equipment to perform the same function in a new location.

Equipment Modification—Major change from original design specifications to an existing unit of equipment or facility.

2

Computerizing Maintenance Management Operations

Salvatore T. Cordaro

Not too many years ago James K. Hildebrand wrote about the negative environment in which maintenance must operate. He wrote:

> We are coping with a precedent that has long placed maintenance at the bottom of the ladder. Besides being a troublesome expense to the thrift-oriented treasurer, maintenance has been the fall guy for many a production head when things go wrong. Maintenance encounters top management only when there is an immediate, serious problem and the atmosphere borders on frantic; or when some large expenditure is necessary which won't contribute anything to real profit. Maintenance therefore, operates in a negative atmosphere. Our greatest achievement is correcting a wrong situation. . . .[1]

For many maintenance organizations, Hildebrand's description of the defensive posture of maintenance is unfortunately as correct today as it was at the time of writing. His suggested solution to the problem of negative image is the computer and the management control it provides. When given the data to make intelligent decisions you can control your environment. Hildebrand advised that the power and organizational ability of the computer would enable the maintenance manager to collect, organize, and report data that would otherwise be too difficult to gather. The validity of Hildebrand's position is especially true for maintenance

---

[1]James K. Hildebrand, *Maintenance Turns to the Computer,* Cahners Publishing Company, Inc., Boston, 1972.

33

since there is typically little clerical support in the maintenance department. With easy and adequate information resources, you will be able to control rather than be controlled. In order to control, however, you must measure and audit.

Until the advent of the microcomputer, Hildebrand's sound advice was difficult to implement. With the growth of microcomputer applications, however, maintenance management systems are now relatively low cost and readily available. You have a wide choice of software because a large volume and variety of "off the shelf," maintenance management systems for the PC are now marketed. Before the arrival of these programs, computerized maintenance was obtainable by only the few larger maintenance organizations. The expense of mainframe and mini-computer systems and the competition for programming resource put smaller maintenance departments on the bottom of the list in most industrial and commercial organizations. On the other hand, the ample budgets of large maintenance departments attracted enough attention from corporate information managers to earn them mainframe time. A number of unique programs specific for the maintenance organization were written. Some of these were later marketed. A typical situation however, was that mainframe and minicomputer systems were almost the exclusive domain of accounting, marketing, production and other related operations. Maintenance with its low priority and status had to wait until the corporate MIS department had the time to develop programs for maintenance applications. The time never seemed to be available because of some new or more pressing priority would arise. This predicament persisted until the microcomputer become a powerful competitor to the mainframe application.

This "breaking of the log jam" with the use of the PC-forced corporate MIS groups to take heed of the needs of maintenance. If they did not, they would be bypassed and lose control of the computer environment. Low cost and the imperative need for improved operations was driving the maintenance manager to computer systems competitive with mini and mainframe applications rigidly controlled by MIS groups. The proliferation of computer software available to satisfy every need of maintenance permits the maintenance organizations to rapidly computerize. And the sophistication of micro software and hardware is increasing almost daily.

FOUR BENEFITS OF USING A COMPUTERIZED MAINTENANCE MANAGEMENT SYSTEM

With a computer-based maintenance management system you can expect improved performance in a number of maintenance activities.

Better Planning and Scheduling of Maintenance Work

Planning time per job decreases, resulting in more plans produced, fewer backlogged jobs, less lost motion by maintenance personnel when they are looking for materials, parts, tools, and documentation.

Job scheduling also becomes more efficient. For example, you will have better access to actual moment-by-moment stores data, thus only jobs with available materials are scheduled. Personnel spend less time waiting and searching for materials. More efficient loading of personnel becomes possible thus reducing dead time between assignments.

More effective work plans can be developed. Due to the resource provided to you by the system, your plans are more detailed and reflect needs realistically. You can have quick recall of repair history, documentation, standard tasks, and procedures. Thus worker job performance and efficiency improve.

Potential Cost Savings: The range of savings from this area can be significant depending on the starting point of the maintenance organization. However, a good rule of thumb is a reduction of 5 to 15 percent of the maintenance labor cost.

Improved Parts Availability

Productivity improvement of the employee is assured through better parts availability. This is effected by better storeroom control. Maintenance personnel typically spend from 12.5 percent to 25 percent of their day searching for parts and materials. A well-organized parts room with user-friendly catalogs that provide easy search features can significantly improve productivity. When storeroom records are inaccurate or nonexistent, a common occurrence is out of stock, lost, or misplaced parts. Personnel lose valuable time in trying to overcome deficiencies. They become idle awaiting new assignments, spend excessive time searching for the missing item, or spend needless time attempting to fabricate a new one or recondition the old one.

Potential Cost Savings: 5 to 10 percent of the maintenance labor cost.

Lower Stores Inventory Holding Costs

The ability of the computer system to maintain moment-by-moment (real time) inventory levels results in a smaller total inventory with fewer stockouts. Computer systems, for example, can provide automatic advice at reorder time, economic order quantities, parts cross referencing, Pareto analysis reports, automatic alert when physical cycle counts are due, quick access to parts usage data, and flagging of nonactive inventory items. Functions such as these contribute to improved stock management efficiency.

Potential Cost Savings: Knowledgeable and tight control of the inventory can mean reduced stock levels. In turn, this can significantly reduce your inventory holding costs. Annual inventory holding costs commonly range from 25 to 35 percent of the value of the inventory. This percentage will vary depending on the value of capital for your company and the type of facility at which the inventory is held. If the above cost of inventory holds true for your facility, it means that for every $1,000,000 held in your maintenance parts and materials storeroom, it can cost your facility from $250,000 to $350,000 each year to maintain that inventory. Good control can mean a 10 to 20 percent reduction in the value of the storeroom inventory. In the first year after reduction, savings will equal 10 to 20 percent of the moneys originally tied up in the inventory. In addition to this one time savings, inventory reduction in the first year will produce lower holding costs. Savings in the following years will only be due to the reduced holding costs or about 25 to 35 percent of the value of the inventory reduced. For example, if your original M & R inventory was $1,000,000, the inventory holding costs are 30 percent, and the inventory reduction is 10 percent or equivalent to $100,000, your savings would be as follows:

First year: $130,000 [(.1 × $1,000,000) + (.3 × $100,000)]
Thereafter: $30,000 (.3 × $100,000)

Increased Machine Availability

Machine operating availability and performance improves because the computer system enables more responsive predictive and preventive maintenance functions. The ability of the system to automatically schedule preventive maintenance puts this valuable program in focus and removes it from neglect. Improved organization and faster retrieval of data and enhanced analysis ability provided by software makes vibration analysis and similar predictive maintenance efforts more powerful than ever.

If properly designed, the system can also improve maintenance engineering applications that focus on isolating and flagging repetitive machine problems. Computer storage of data regarding repair history and failure permits faster response time by maintenance engineering personnel. Armed with the knowledge of which defects are the most common problems, the maintenance engineer and others working on improving equipment availability can design retro-fits and refinements which eliminate the weakness which cause failure. The mean life of the equipment can be extended and maintainability can be improved.

Potential Cost Savings: The range of saving for this activity can vary considerably depending on the current philosophy of the maintenance organization. If your organization is in a "fix it when it breaks" mode of operations, not much can be saved until the correct organizational view takes hold. Savings can range from as little as 0.5 percent increase in up-time, to as much as 15 percent. Savings in these applications can be quite significant and can exceed the savings in all other areas.

DETERMINING YOUR MAINTENANCE COMPUTER SYSTEM REQUIREMENTS

As your organizational needs grow, a simple standalone PC may not do the job. In most maintenance departments several work stations will be required to handle the input and service the needs for various areas of responsibility. For example, one station may be required for the maintenance storeroom, one or several may be required for a planning group, one may be required by the maintenance manager, one for the first line supervisor, etc. If the organizational culture has progressed to worker involvement, greater emphasis will be placed on the hourly team member interfacing with the system. In this environment, perhaps several additional work stations will be required on the shop floor and other key locations to service the teams.

In the near past, multiuser requirements of this nature demanded mini, or mainframe platforms. Today the local area network (LAN), which has matured significantly in the past several years, can do the job. The LAN can sometimes provide more power at lower cost than some mini systems.

PLANNING THE INSTALLATION OF THE MAINTENANCE COMPUTER SYSTEM

It is an unfortunate fact that many maintenance organizations underestimate the time required to install their database. The purchase of software and hardware is only the starting point of a difficult process of installation and start up. Before you can make a new system functional, you must identify plant assets, preventive maintenance programs, inventory items, and other information comprising a large body of data. You should then enter this data into the system files.

This process becomes especially difficult if your maintenance organization does not already have this information in some format such as in a manual system. If the data is not already available, you must embark on a major project of inventory and identification. For example, you should inventory all assets which will be included in a preventive maintenance program, or for which work orders or repair history will be generated. Identify the assets by description, manufacturer, manufacturer's model number, and a unique number for the facility which is commonly referred to as the equipment number.

Most software systems also permit the entry of related asset information which is useful for quick reference. For example, warranty information, drawing information, serial number, capacity, physical location, accounting codes, department number, spare parts lists, and other details must be entered into the computer data base that is stored. This type of detail can be important when you or your staff are planning repair jobs or are generating work orders. The task is large when you consider that this type of information must be collected, verified correct, and then keyed into the system.

In the typical maintenance organization, some if not many, of the key manufacturer's manuals are often missing. Thus, your task may also include an inventory of the source documents and the pursuit and acquisition of the missing documentation.

Similar efforts must be generated for each stores inventory item if the storeroom is to be included in the computer system. The inventory item must be identified by manufacturer, manufacturer part number, description, site inventory number, bin address in the storeroom, machine which it spares, and other details.

Vendor data must also be collected if the storeroom system interfaces with purchasing. If you are starting with little prior information, you may have a one-to-two year project on your hands, depending on the people resources available to do the work.

If you have PM software, preventive maintenance tasks must be defined, tasks scheduled, and task material and parts lists developed. All this information must then be keyed into the system.

If you fail to assign sufficient resources in terms of people on this data collection, purification, and entry, you will find two or more years will go by before your system becomes functional and begins to provide the return it promises. To avoid this, budgets for the computer project should include costs of system start-up. These costs should include the salaries of those who will devote their time to the computer implementation project. Include sufficient people resources for collection and input of the computer data base, and the start-up of the system. These efforts will constitute full-time jobs when the process begins and should not be relegated to a part-time department assignment.

Some costs, especially those associated with hardware, can be anticipated with some precision. Software costs can be targeted and will generally reflect the sophistication of the systems desired. Unfortunately start-up costs cannot be predicted with any absolute assurance. They should therefore receive a great deal of attention in project development. The architecture of the program that is finally selected will have a significant bearing on these costs and should be carefully defined.

Just how many people will be required to install the system's database? Again this cannot be judged accurately unless specific information is available in terms of numbers of items and status of records. A typical experience for a small department with perhaps 30 to 40 maintenance people, however, might provide a guide. Experience has shown that it requires eight months to one year with two employees working full time, to collect and input the data necessary to activate all the features of a full spectrum maintenance management system. Your resource needs will vary considerably on the state of your current documentation. This variable must be carefully considered when budgeting cost and planning start-up.

You should also consider the cost of training personnel in the methods of the new computer system. It has been estimated that the cost of training is approximately 50 percent of the cost of the software.

SIX MAINTENANCE ACTIVITIES THAT CAN BE IMPROVED WITH COMPUTERS

Maintenance computer systems can improve maintenance activities in the following ways:

- improve procurement and control of the spare parts inventory
- facilitate predictive and preventive maintenance activities

- record information on equipment specifications and repair history and repair costs
- improve work order control, planning, and scheduling
- facilitate purchasing of materials and parts
- generate reports that measure the performance of the maintenance function

These six improvements are discussed in detail in the following sections.

How to Improve Procurement and Control of the Spare Parts Inventory

The system should provide you a spare parts catalog indexed for ease of information retrieval by maintenance workers. It should permit the planner to determine on-line the availability of parts and materials and provide inventory control so that stock-outs are eliminated. It should maintain accounting information internally within the system by transferring parts cost to the maintenance work order against which the part was issued. Specifically, the system should perform the following functions:

- Support a spare parts catalog indexed for ease of information retrieval by your maintenance workers. Typically, they search for a part by generic name, manufacturer's part number, or by the equipment number for which the part is a spare. They should be able to solicit and obtain information by any one of these three reference methods. Once having identified the item, the system should provide as a minimum, the following information:
  - catalog description of the part,
  - catalog (facility) inventory number,
  - the address in the storeroom where the part is located,
  - the quantity currently on hand.

- Permit planners to determine on-line whether parts and materials for a scheduled job are available. In some applications it may also be desirable for the planner to be able to reserve inventory items against a scheduled job.

- Provide an inventory control system so that stockouts are eliminated. The system should respond with an immediate, on-line balance on hand, when inventory items are issued or received. It should also alert the storekeeper when a reorder point is reached and indicate "on-order" quantity when the order is placed. When ready to order, the system should provide your stores clerk with information on reorder quantity, preferred vendor, lead time, shelf life, and unit price.

  Items due for a cyclic physical inventory should also be flagged so that each day a portion of the inventory may be checked by actual count in order to verify system "balance on hand."

- Report management information such as:

- usage,
- inactive items,
- total inventory value by item ranked in order of dollar value,
- physical inventory variance from system count,
- cost of issued items by cost center or work order number

Other useful functions you should consider for the storeroom application include system ability to print bin labels and bar code applications.

How to Facilitate Predictive and Preventive Maintenance

You should seek software containing predictive maintenance features. The measurement of equipment condition is now a commonplace method for ensuring against breakdown. Vibration, temperature, wear particle accumulation in oil systems, oil contamination, ultrasonic sound absorption, all help show the state of health of a unit. These methods, when applicable, are in many ways superior to the traditional methods of equipment surveillance usually called preventive maintenance.

Your system should include features that provide the functions described in the following paragraphs.

- *The system should provide methods for scheduling repetitive predictive maintenance data collection tours of specified equipment for oil samples, infrared temperature readings, and so on.* For example, in a vibration program, a weekly tour might be required which takes the vibration data collection person through the facility in a logical sequence. The system identifies the machines that are monitored, the vibration data collection points on the machine to be collected, and the vibration limits above which some corrective action must be taken. This information is printed to a schedule sheet. The schedule sheet also should contain space for recording instrument readings. Upon completion of the tour, data recorded is input to the system that stores the data and generates reports. For example, an exception report might list those collection points which have exceeded some alarm limit. In this manner, you can quickly investigate and correct an anomaly that could lead to a breakdown.

  Very sophisticated, specialized software for vibration analysis programs are also available through the vibration instrument supplier. These systems are very powerful and can outdistance what is available through normal maintenance management systems. These specialized vibration software programs are normally associated with specialized vibration data collectors. The automated data collector is a portable vibration instrument that houses its own computer system within the instrument case. The data collector can store routes, machine identification, vibration check points on the machine, vibration limits, and so on, and the measured vibration data collected during the route travelled. At the start of the work day, the user connects the data collector to a host computer and down-loads data to it. Vibration data collected is uploaded to the host at the completion of the route. Then the host with its specialized vibration software will store, analyze, and report the information collected. These are excellent systems and should be part of an overall equipment reliability effort.

Vibration applications that are part of a maintenance management software package are not as sophisticated as the above. These simpler systems, however, can be useful for some types of maintenance environments. When the organization uses simpler vibration instruments that depend on manual collection of the data, these systems work well. These systems also are adaptable to other types of predictive data collection routes; i.e., oil sample collection routes.

- *Your system should provide information that highlights your major causes of equipment breakdown and unscheduled repairs.* It should be able to collect and report defects that cause failure by type and frequency. Through various predefined codes, a history of the types of problems that lead to breakdown and maintenance repair can be collected. Pareto-type reports can then help direct maintenance engineering efforts aimed at reducing the incidence of failure and weakness in equipment. These techniques work because relatively few defects (20%) cause the most maintenance problems (80%). If you identify and document these important defect types, you can redesign and retrofit to reduce the incidence of failure and the requirements for maintenance work.

- *Software permits the generation of preventive maintenance tasks that, when due, become part of the preventive maintenance work order.* Computer applications are ideal for preventive maintenance surveillance methods. The traditional preventive maintenance functions based on timed inspections are still an important and principal activity for any maintenance effort. Software systems can store task descriptions and release preventive maintenance work orders (orders for PM inspections) at intervals defined by the preventive maintenance program.

The system should permit the compilation of tasks which are either generic for many units, or are specific for a single unit. The file containing these tasks may be considered a library of tasks, from which can be drawn work elements comprising any type of work order whether it is PM or repair.

The system you select should permit the scheduling of maintenance tasks against equipment either on a calendar basis or by hours of operation. The task defined should be linked to a listing of materials and parts required for the execution of the task. When the preventive work order is generated, a bill of materials is associated with it so that the person doing the PM knows the materials requirements before going to the asset. The skills requirements, number of persons per task, and total hours to execute the work are also important features that must be included in the PM work order.

To facilitate the setup of the preventive maintenance system, a desirable feature would be one which permits the copying of tasks and/or materials requirements to other equipment. This feature would be utilized when equipment is identical or similar to the first one for which a PM program has been defined.

Your system should generate a report of projected PM labor requirements by craft. Important report functions include the ability to project labor requirements for the PM work orders over a span of weeks or months. At a requested starting date, the report compiles weekly labor requirements produced by the PM schedule for a requested projection of time. A reasonable span of time would be 13 weeks, which is a quarterly projection. Reporting of backlog PM work orders is also an important feature for control of the program. Overdue work orders must be pursued to assure compliance with the system. A print of all tasks that have been defined in the library is also important since this print would constitute the PM catalog.

How to Record Information on Equipment Specifications and Repair History and Costs

Additional software applications which you should require in your system are those which deal with equipment data and specifications. The equipment specification is one of several primary hubs in the software system. Information essential to planning, getting repair parts and executing maintenance work should be kept in your software files.

Keep a Record of Equipment Specifications. Detailed nameplate and other supplemental data should be kept in these files for ready reference. These equipment files provide a form of rapid information to those who need to plan work orders, order materials and parts for repair applications, or identify some item of information unique to the machine such as a serial number.

Maintain Files on Equipment Repair History. This information is obviously useful to both the maintenance manager who wishes to review the failure record of the equipment and the planner who must design scope of work for new repair activities. Repair history may be kept at two levels: a cost level and a "what was done" level. Each has functional value and should be considered an important feature when searching for software systems.

This application might also be the location for storing "model" work orders. These are work orders that have been executed, but are typical of many similar jobs. They provide a benchmark and quick reference for the planner. If similar work is planned, model work orders can be called up, modified to fit the specifics of the job planned, and then issued. This can save considerable planning time since the model work order will have contained estimates of labor, the various tasks required for job execution, materials and parts requirements and identification, and special tools that may be required.

Generate and report budgets by unit and/or by cost center. The system should be designed to generate reports which compare budget versus actual maintenance expenditures for equipment and/or cost centers. A field in these reports would display the variance from budget.

Associated reports would include a detail of the work orders that were executed and which make up the repair costs. For example, a detail report would list the work orders by work order number and work order title. These would be orders executed in the period reported. The print would display labor, material, and contract costs contributing to actual maintenance costs for the equipment. With such information, one may search the work order files to determine the specific causes for budget overruns.

How to Improve Work Order Control, Planning, and Scheduling

Probably the most important applications for improving work execution performance are those that generate, track and help process work orders and schedules. These systems would also generally be the place where work order repair history, actual labor hours, closeout, and so on, are keyed to the system. The program should provide a screen display for entry of the work request, a function for storing these requests, and a method which permits display of the request at any time for approval and conversion into a work order.

The system should provide data for the planner to determine scope of the job, providing a method for listing in sequence the specific tasks that make up the work required, and estimate the skills and labor requirements. It should also provide support for determining and listing part requirements and availability and special tools, and the like. The system should also track work orders by status and permit scheduling on both a weekly and daily basis.

The system should permit the entry of maintenance work requests. The screen display for this type of feature should be in color and easy to read, since request entry is often by the customer who does not wish to expend excessive time in this effort. A fill-in-the-blanks format can make the task easy for just about everyone. In some formats, the request is just as the name implies, a request. (Comparison can be made with the purchase request originated by an employee and submitted to the purchasing department.)

Using a terminal and screen display, an authorized maintenance person (i.e., the planner) reviews the request and converts it into a work order. (The comparison being a purchase order is generated by the purchasing department after assurance the request meets organization approval criterion.) The system permits the sorting, selection and display of the work request for approval.

The System should facilitate Work Order Planning. The work order is then planned utilizing features of the system. The planner:

• determines scope (using the repair history features),

- phases the job into a sequence of tasks (when appropriate, using a library of tasks i.e., bench mark tasks and/or model work orders, that are stored in the system),
- determines the skills requirement and the number of persons required to execute each task and lists these on the work order,
- determines and lists the materials and tools requirements.

The system provides ready reference to the planner so he or she may speedily accomplish the planning tasks and quickly identify and determine availability of the materials, parts, special tools and documentation to be used in execution of the job.

The system should permit the generation of a weekly and daily maintenance schedule. The schedule is compiled according to priority and availability of labor by craft. Preventive maintenance work orders are conveniently scheduled using this system.

The system should permit the tracking of maintenance work orders and their status. For example, backlog, released, and overdue work orders are reported. Search features permit the sorting of work by specified classifications. For example, all work identified as "shut down" work is filtered and reported two weeks prior to a scheduled shut down. This permits the planner to integrate these jobs with the shut down work.

The system should provide entry screens to permit the recording of completed maintenance work, actual hours worked, type of defect encountered and downtime hours experienced. Material and parts costs are automatically recorded when they are issued against the work order number through the "storeroom" module.

The system should permit the assignment of objective work priorities. The system priority mechanism should consider the importance of the asset which requires maintenance work and the type of work to be performed.

The system should permit the compilation of standard maintenance tasks in a file of tasks. When preventive maintenance or any deferrable work is planned, the user can utilize these task by "pasting" them to the work order. The tasks can consist of any task previously designed and saved. They can also include formal job standards, safety procedures or standard procedures designed by the maintenance organization for certain types of work to be performed in the facility.

It would also be desirable for the application to have a method for saving previously generated work orders to be used as model work orders. These orders, although generated in the execution of specific jobs, would have a generic character which would lend itself for quickly designing work orders for many similar jobs that arise from time to time.

How to Facilitate Purchasing of Materials and Parts

When maintenance personnel are involved in purchasing materials and parts, a purchasing function is helpful. Examples of such activity would be the replenishment of storeroom stock, the planner ordering non-stock items for a repair job, emergency purchases required because of a breakdown.

The software purchasing application should provide the user with a list of vendors who supply the particular inventory item to be ordered. The function should be designed to automatically paste the vendor information into the purchase request when it is generated.

Many programs will automatically generate a purchase requisition when a re-order point for an inventory item is reached. The advantage of the automatic feature is, of course, the elimination of the time needed to prepare the requisition. In some circumstances however, the automatic generation of a request may have a negative side, i.e., the bypassing of intervention by some knowledgeable person who is familiar with the operation and can make judgments that override an automatic system. For example, a user may decide not to order at the re-order point because of an impending plant vacation, or he may wish to order more than the usual order quantity because the facility is entering a turnaround period.

Good purchasing systems will also permit the tracking of vendor performance so that the quality of vendor service may be documented. Lead time, number of items backlogged, late deliveries, goods returned, and so on, could be valuable information.

How to Generate Reports That Help Control Maintenance Performance

Your maintenance system should be able to issue a number of different types of reports which measure the performance of the maintenance operation. The reports should be daily, weekly, monthly semi-annually and perhaps less frequently depending on needs. The reports should also be formatted to accommodate the different levels of management. For example, the department manager will receive summary information, whereas the first line supervisor will receive information specific to his or her immediate area of responsibility.

Your system should be able to report on the following factors:

- Backlog Trends—The system should provide tables or preferably graphs showing backlog in weeks for each craft.
- Performance—Records should indicate the labor hours estimated versus the actual labor hours required to execute the work order. This report could be cumulative by the week for craft, shop, or department.
- Work Order Accounting—The system should track the labor, material and contracting costs for each job. Cumulative costs by equipment or cost center would also be important.
- Defect Analysis—Downtime statistics by unit are important performance and maintenance engineering data. This data would further contain information on the problem, problem cause and method of correction. The frequency of occurrence and the percentage of downtime caused by each problem group is also important.

- Value of Maintenance Inventory—Total value by descending order of contribution to the inventory dollar value is an important ABC analysis report which can help control stocking levels. Other reports useful in the storeroom are activity reports, slow-moving inventory reports which flag candidates for obsolescence, variance from physical counts reports, and stock out reports.

- Conformance to Schedule—Important information for the planning function would be the percentage of the schedule completed, the allocation of labor to types of maintenance, i.e., emergency, planned and scheduled, and the percentage of materials requirements anticipated and coordinated by the planners.

COMPILING, ORGANIZING, VERIFYING, AND INPUTTING INFORMATION ON EQUIPMENT MAINTENANCE

If you haven't allowed enough money in your project budget for this fundamental activity of database building, you can expect serious delays in system implementation. The variety and amount of data which must be collected and verified makes this job almost overwhelming. Organizations have required two or more years in completion of this phase because of inadequate personnel budgeting. Can your project afford such delays before it begins to provide its promised return?

The task becomes especially monumental if you are in transition from a paper system and much of the data required must be assembled for the first time. Hopefully, you already have much of this information since this is essential data whether or not you turn to computer management systems.

A caution however: If you are generating equipment and inventory numbers for the first time, you need to be concerned with field sizes. Since you are going to use a computer system, the numbers you design do not have to be "intelligent" numbers—that is, the number does not have to carry any imbedded descriptive code number such as for a commodity or equipment type. The computer system will have the ability to search, inform, and help define such descriptive elements which have been traditionally been part of paper system numbers. This means you may and should keep your numbers as simple as possible. The numbers you design should be only numeric, preferably of six or less characters, without spaces, alpha characters or dashes. If you design numbers of this type you may be assured that they will match field sizes defined by most software packages available. Such a design will also help avoid errors and increase input speed.

Your computer data base should include the following information:

- an inventory of facility equipment,
- a unique equipment number used to identify each unit,
- a file containing unit name plate data and specifications which has been verified by field confirmation,

- a list containing each part and maintenance material item to be kept in inventory including necessary information on these items such as noun identifier, description, manufacturer's name, manufacturer's part number, facility stock number, storeroom address, reorder point, reorder quantity, other information such as vendor information which may be contained in a software system,
- a manual of preventive maintenance tasks and frequency of inspection for each item of equipment which will be included in the PM program,
- a library of manufacturer's repair manuals, schematics, drawings, etc., which, depending on the sophistication of the software system, will provide the hard copy for laser disc and cad-cam reference files.

Getting data into the system is another major effort. After collecting the basic documentation, it is a good idea to record the information that is to be transferred to the computer system on to input forms. If data is to be entered via a terminal, the source documents must be designed with the interactive dialogue in mind. These forms should be designed to match the size and sequence of the fields of information contained in the system's input format on the display screen.

The following flowcharts show applications that are available in a currently marketed, canned, microcomputer-based maintenance management system. This system contains six major elements that cover the range of maintenance management applications. These elements bear the names, in order of presentation, *STORES/2, SWOP, PREVENT/2, EQUIPMENTS, PDM,* and *PURCHASE/2.* In addition, the software contains a user generated report writer. You may use system elements as stand alone units or as an integrated software program for complete direction of maintenance activities. *SWOP* is the name given the module for work order generation, tracking, planning, and scheduling. The other software units cover activities implied by their names.

Figures 2-1 through 2-4 show applications available with the maintenance storeroom software module.

Figures 2-5 through 2-7, display applications available for work order control, planning, and scheduling.

Figure 2-8 illustrates functions available with the module devoted to preventive maintenance applications. When this application generates a PM work order, the *SWOP* module picks it up for scheduling.

Figures 2-9 and 2-10 show equipment module applications. This module provides technical specifications and repair records on equipment. Another significant application provided by this software element is equipment repair cost and budgeting.

Figure 2-11 shows a flowchart of predictive maintenance applications and *Figure 2-12* shows system support for maintenance purchasing functions.

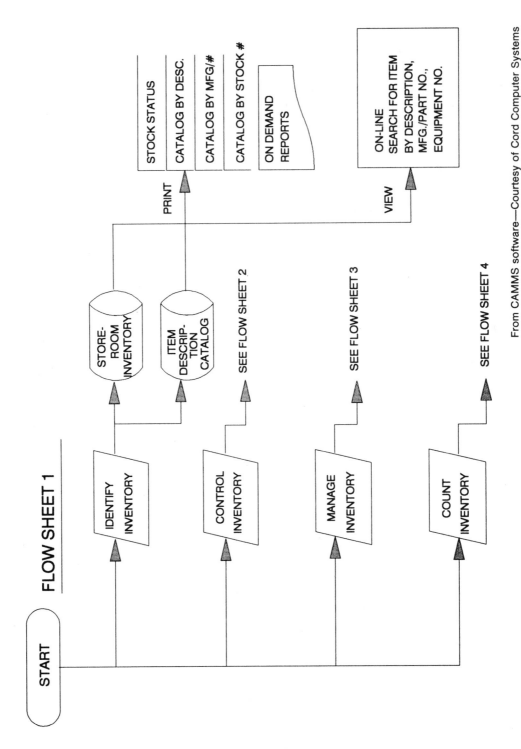

Figure 2-1. Flow Sheet Illustrating Major Stores Activities.

From CAMMS software—Courtesy of Cord Computer Systems

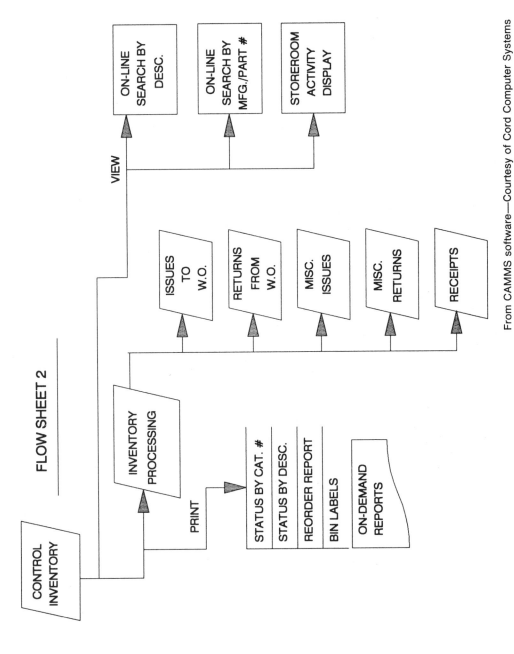

Figure 2-2. Flow Sheet 2—Simplified Illustration of Storeroom Control Applications

From CAMMS software—Courtesy of Cord Computer Systems

49

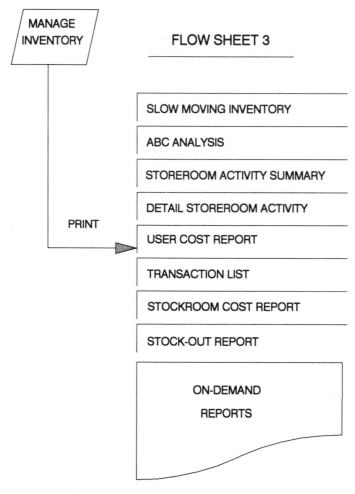

Figure 2-3. Flow Sheet 3—Showing Some of the Reports Generated for Storeroom Management

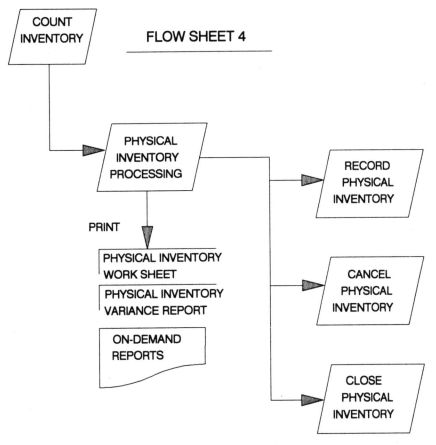

From CAMMS software—Courtesy of Cord Computer Systems

Figure 2-4. Flow Sheet 4—Showing Physical Inventory Applications

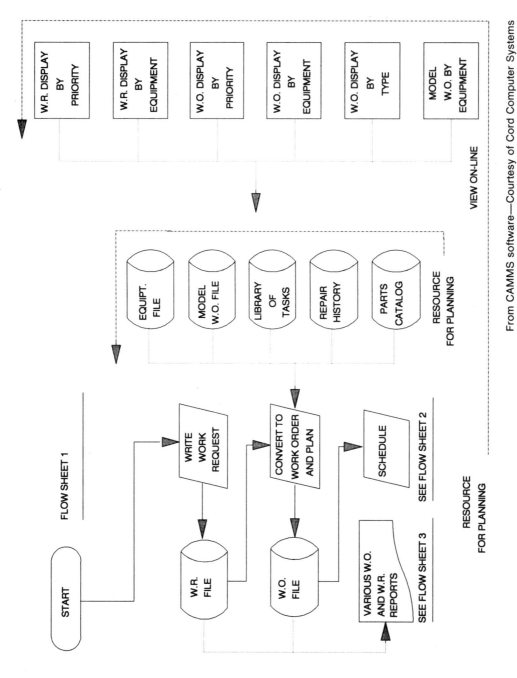

Figure 2-5. Flow Sheet Showing Activities in the Generation, Tracking and Planning of Work Orders

From CAMMS software—Courtesy of Cord Computer Systems

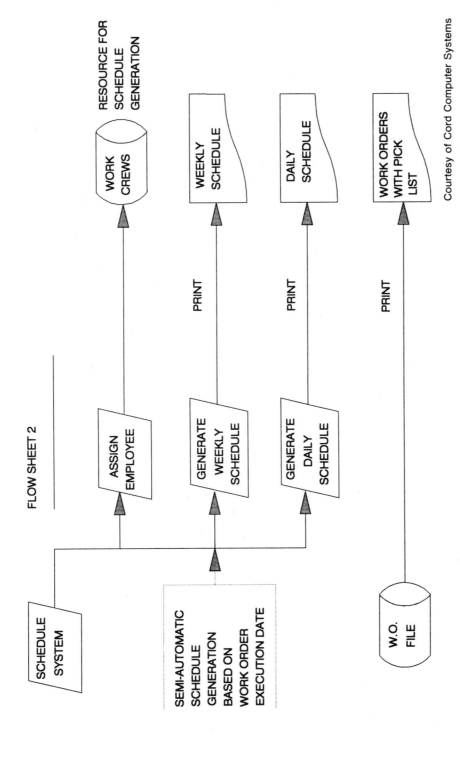

Figure 2-6. Flow Sheet 2 Showing a Simplified Detail of the CAMMS Application for Scheduling Work Orders

Courtesy of Cord Computer Systems

53

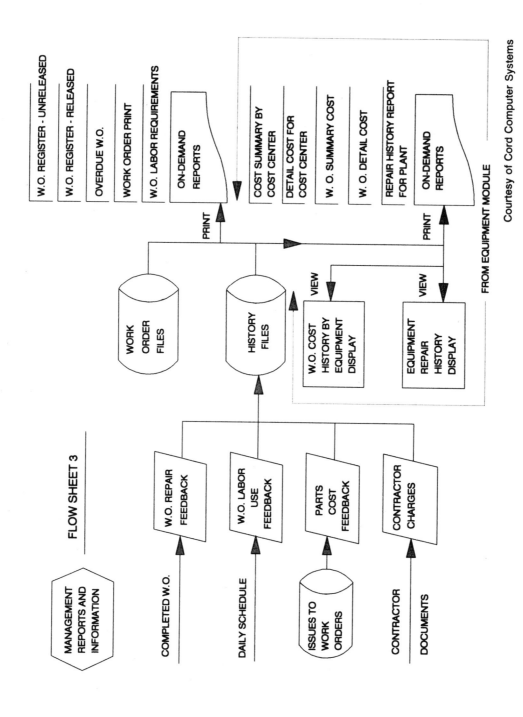

Figure 2-7. Flow Sheet 3 Showing Simplified Detail of Management Reports and Information Available in the SWOP Module of CAMMS

Courtesy of Cord Computer Systems

54

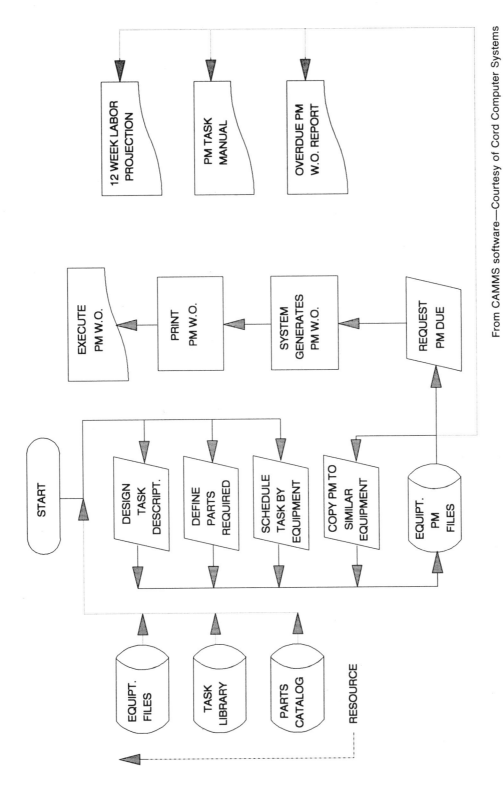

Figure 2-8. Flow Sheet Illustrating Major Preventive Maintenance Activities

From CAMMS software—Courtesy of Cord Computer Systems

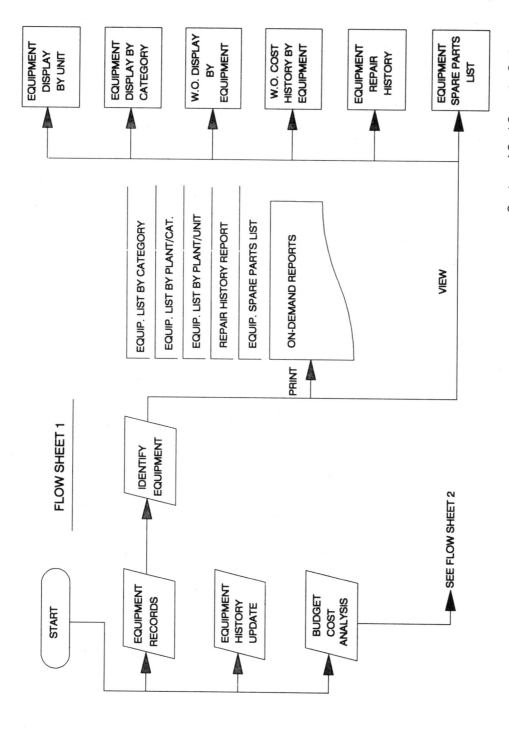

Figure 2-9. Simplified Flow Sheet Showing Activities Available in the EQUIPMENTS Module of the CAMMS Software

Courtesy of Cord Computer Systems

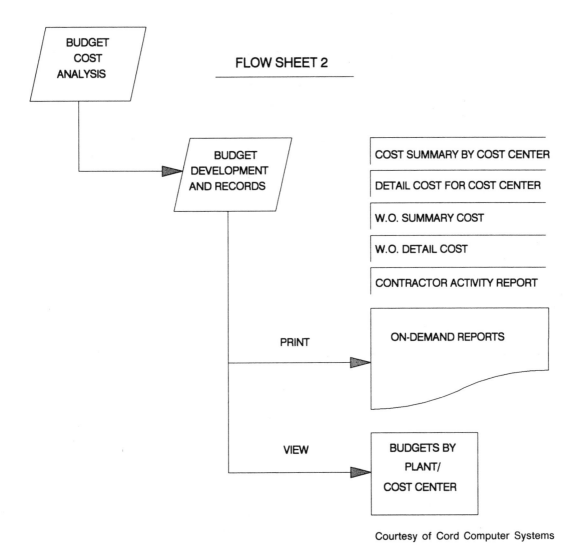

FLOW SHEET 2

BUDGET COST ANALYSIS

BUDGET DEVELOPMENT AND RECORDS

COST SUMMARY BY COST CENTER

DETAIL COST FOR COST CENTER

W.O. SUMMARY COST

W.O. DETAIL COST

CONTRACTOR ACTIVITY REPORT

PRINT

ON-DEMAND REPORTS

VIEW

BUDGETS BY PLANT/ COST CENTER

Courtesy of Cord Computer Systems

Figure 2-10. Flow Sheet 2 of Budgeting Activities Available Through the EQUIPMENTS Module of the CAMMS Software

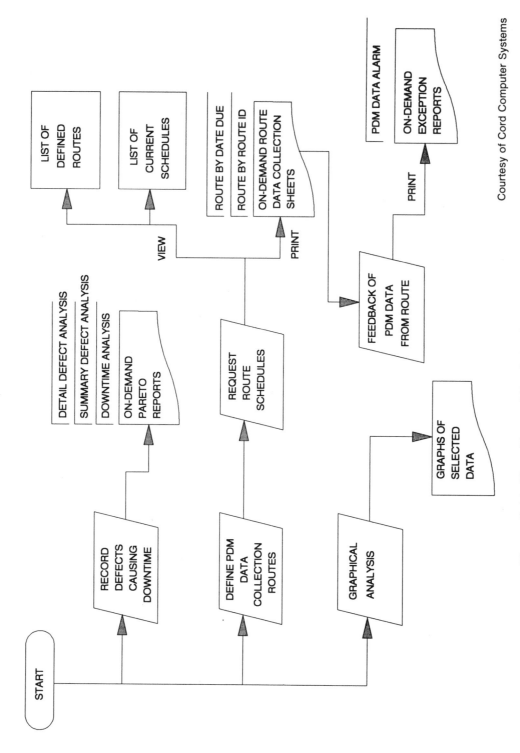

Figure 2-11. Flow Sheet of Predictive Maintenance Applications Available in CAMMS Software

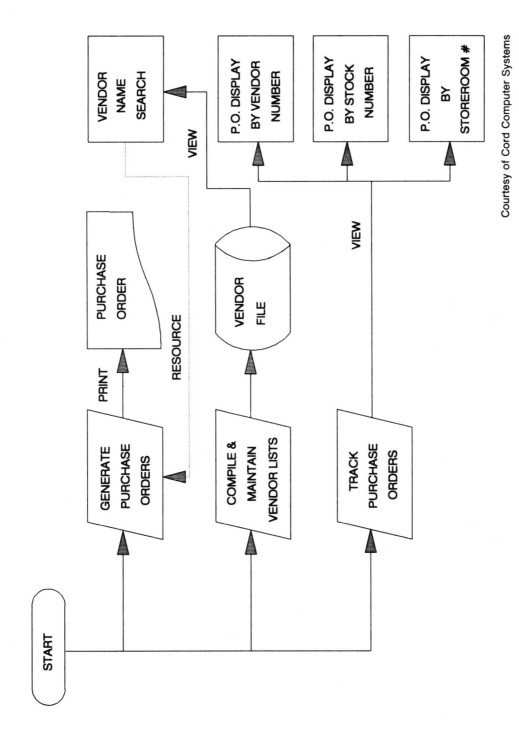

Figure 2-12. Flow Sheet Showing the Major Features of the Purchasing Applications in CAMMS Software

Courtesy of Cord Computer Systems

3

Lighting System Maintenance

Robert Nigrello

Good lighting is an inseparable part of our modern life as it has been for decades. It is vital to business, industry, and commerce. It provides safety and security, aids our work, our pleasures and helps us achieve our goals.

Well-maintained, efficient lighting speeds production, improves product quality, reduces errors, spoilage, and accidents for industry. It also enhances teaching and learning in schools and eases routines, increases efficiency, and helps the comfort of personnel in offices.

Good lighting means lighting maintenance that makes economic and good business sense. It ties directly into preserving our national energy supplies and keeps the cost of electrical power to the minimum.

DEFINING LIGHTING SYSTEMS

Rated Lamp Life

Because of variations in lamp-making operations and lamp materials, it is impossible to have a lamp operate for exactly the life for which it was designed. For this reason lamp life is rated as the average life of a large group of lamps. For example, a group of 100 incandescent lamps rated at 750 hours may have 50 lamps still burning at the 750 hour mark; 50 lamps would have expired during that time.

For ballasted lamps, rated life is a bit more complicated. For example, a fluorescent lamp may be rated at 20,000 hours. This is only for lamps burned for three hours per start. Longer burning times will increase average rated life. Shorter burning times—of less than three hours-will decrease average life. Table 3-1 shows fluorescent lamp life for extended burning periods.

Lamp Type	Hours Per Start					
	3	6	10	12	18	Continuous
40W Preheat	15,000	17,500	21,250	22,500	25,000	28,125
40W Rapid Start	20,000 +	24,420	27,750	28,860	31,600	37,700
High Output (HO)	12,000	14,000	17,000	18,000	20,000	22,500
Very High Output (VHO)	10,000	12,500	14,990	15,980	17,980	24,980
Slimline (96T 12)	12,000	14,000	17,000	18,000	20,000	22,500

Courtesy of Osram Sylvania Lighting

Table 3-1. Average Life in Hours of Fluorescent Lamps at Various Burning Cycles

Color Temperature (CCT)

The color temperature of a light source is expressed in Kelvins (K). Sources having a low color temperature (2700K to 3350K) are said to be warm. Mid range colors are centered at 3500K. Cool colors are at 4000K and above. The color temperature of a lamp contributes greatly to the visual appearance of a space. A typical example is the warm red appearance of an incandescent lamp. This incandescent lamp is approximately 2700K. On the other extreme is the blue sky on a bright clear day. This cool blue is approximately 6300K.

Color Rendering Index (CRI)

The Color Rendering Index is an international numbering system and ranges from 0-100; it indicates relative color properties of a light source, and is abbreviated CRI. The higher the number, the better the color rendering properties of the source will be. Incandescent lamps have a CRI of 97. This makes skin tones and most colors look vibrant. Cool white fluorescent lamps are rated at 62 CRI. This low number causes colors to look flat and skin tones to look pale. In general, a lamp with a CRI of 70 and above will render colors acceptable; below 70 CRI, colors will look flat. A rather interesting lamp is the low pressure sodium lamp. The CRI of this source is 0. This translates into a lamp which distorts colors. For example, under low pressure sodium a red, yellow, or blue colored car will all look a shade of brown.

Efficiency (LPW)

Lumen per watt describes the efficiency of the light source much like miles per gallon describes auto efficiency. The least efficient source is incandescent at 20 LPW and the most efficient source is low pressure sodium at 180 LPW. Table 3-2 lists the LPW, CRI, and color temperature of the most common light sources.

LAMP TYPE	LPW	CRI	COLOR TEMPERATURE
INCANDESCENT	17.5	97	2700-3200K
TUNGSTEN HALOGEN	20	97	3000-3200K
FLUORESCENT	55-100	50-90	2700-7500K
METAL ARC	70-100	65-85	3000-5000K
MERCURY	40-60	22-55	3000-5900K
HIGH PRESSURE SODIUM	50-140	22-80	1800-2600K
LOW PRESSURE SODIUM	100-180	0	1800K

Table 3-2. LPW, CRI, and CCT of Lamp Families

IDENTIFYING LIGHTING SYSTEM COMPONENTS

Ballast

A ballast is a current limiting device that provides the proper voltage and current to discharge (fluorescent and HID) lamps. It may also provide a starting pulse for HID lamps. Ballasts can be constructed of core and coil (magnetic) or electronic components.

Fluorescent Ballasts

Preheat Ballast

The preheat ballast was the first fluorescent ballast type available. It was developed in the 1930s. In addition to the ballast, a starter is needed to start a fluorescent lamp. A distinct feature of this type of ballast is the way it starts a lamp. The characteristic blink-blink-blink before the lamp starts, tells the maintenance staff that they are dealing with a preheat ballast. Although still manufactured for task lighting and smaller lamps, preheat ballasts are rarely found in commercial or industrial applications. There is an exception in compact fluorescent lamps. This fluorescent lamp family uses primarily preheat ballasts.

Instant Start Ballast

The instant start ballast, as its name implies, instantly starts a fluorescent lamp. It is always used with eight-foot F96 type lamps that do not carry a HO or

VHO designation. All lamps that use this type of ballast have a single pin on each end of the lamp. This contrasts with preheat and rapid start systems which have a bi-pin base.

Rapid Start Ballast

The rapid start ballast is by far the most common type of fluorescent system ballast. Its starting time is faster than preheat but not as fast as instant start. This type of ballast contributes to long lamp life of over 20,000 hours on four-foot lamps . The characteristic bi-pin base is found on most rapid start lamps. Also included in rapid start circuits are High Output (HO) and Very High Output (VHO) lamps. This family of fluorescents has a recessed double contact (RDC) base that is easily identifiable to Maintenance personnel.

Fluorescent Ballast Efficiency

Federal law requires fluorescent ballast to exhibit a minimum efficiency. No ballast manufacturer can manufacture a ballast for use in the United States that does not meet these minimum requirements. However, more efficient ballasts are available. A fluorescent ballast that is electronic is a more efficient device. These ballasts deliver current to the lamp at much higher frequency than magnetic ballasts. The mercury within the lamp uses this high-frequency current to produce the same light at less wattage. Typically, a 30 percent reduction in wattage is obtained by using an electronic ballast.

Electronic fluorescent ballasts have been available since the late seventies, but it is only since the mid-eighties that they have earned a reputation for high reliability.

High Intensity Discharge Ballasts

Reactor Ballast

A reactor ballast is the simplest type of HID ballast, commonly called a choke. Lamp wattage varies by a 5 percent change in input voltage and causes a 12 percent change in lamp current. A voltage drop of 25 percent of rated input voltage will cause the lamp to extinguish. This type of ballast is commonly used in Mercury lighting systems.

Constant Wattage Autotransformer

The constant wattage autotransformer, commonly called the CWA, is the most common HID ballast used because it offers the best compromise between cost and performance. A 10 percent change in voltage results in only a 5 percent change in lamp wattage. For this ballast a decrease in voltage of 30-40 percent is needed before the lamp will extinguish. Mercury, Sodium, and Metal Arc lamps can all be powered by this ballast.

Constant Wattage

The constant wattage (CW) type of ballast provides the highest wattage regulation. Ballasts of this type can have a 13 percent change in voltage with only a 2 percent change in system wattage. Because there is no connection between the primary and secondary, this ballast is also an isolated circuit providing the extra security an isolated circuit provides. All HID systems can use a CW ballast.

Lighting Fixture

Any enclosure that emits light in a controlled pattern can be called a fixture. This can be as simple as a socket with an incandescent lamp or a one-of-a-kind crystal chandelier designed by a world-renowned designer.

Incandescent and tungsten halogen fixtures are available for all applications, from downlighting to floodlighting. The reflector design controls the way light is distributed for all nonreflector type lamps. Trim and louver selection further define the projected light beam. Since incandescent lamps are a line voltage source, ballasts are not contained in these fixtures. A transformer will be present for a lamp whose rated voltage is different from the available line voltage. An example of a transformer system is low voltage lighting. This is a 12-volt system, very common in tungsten halogen MR 16 type lamps.

Fixtures that use reflector lamps rely on the lamp reflector to control the beam. These fixtures do not contain internal reflectors.

Fluorescent fixtures are commonly configured in 2×2, 2×4 and 1×4's containing one, two, three, or four lamps. Due to the nature of fluorescent lamps, a ballast will be located in the fixture. These "troffers" direct fluorescent light by directing it off the rear reflector and through the front lens or louver. Any modification of reflector or lens will alter the light distribution. Down light cans containing compact fluorescent are very popular because of their efficiency over incandescent lamps.

HID fixtures are used for interior and exterior lighting. They are available in High Bay, Low Bay, Recessed Cans, Floodlighting, and Uplights for indirect lighting. HID lighting fixtures control light by using reflectors or multiple reflectors. Lenses on HID fixtures are usually not for light control but are important for safety and ultra violet control. A ballast is always contained in an HID fixture because it is required for powering an HID lamp.

Lamps

Lamps are commonly called light bulb lamps and come in a wide variety of shapes and sizes. Lamps are divided into three categories: Incandescent, Fluorescent, and HID.

Each group has many common elements and are described below. Bulb shape describes the physical shape of the lamp. Some of the more common shapes are A, Globe, Pear Shape, and Tubular. The number associated with the shape indicates the maximum diameter of the lamp in eighths of inches. For example an A19 lamp is an A shape lamp with a diameter of 19/8 or 2 3/8 inches. Another example is a T12 lamp. This would be a tubular lamp 12/8 inches or 1 1/2 inches in diameter. Lighted length (LL) is the distance in inches from the base of the lamp to the middle of the lighted element. Mean overall length (MOL) is the total length in inches of the lamp. The base of the lamp can be medium, mogul, recessed double contact, and bi-pin to mention only a few of the many lamp bases. Figure 3-1 shows in detail the various lamp parts and the basics of a lamp ordering guide. (For unusual circumstances, OSRAM Sylvania lighting will identify lamps at the following number: 1-800 255 5042.)

MINIATURE LAMPS

The term "miniature lamp" was developed as a convention early in the history of the lighting industry. It actually applies to units ranging in size from the so-called "grain of wheat" lamps to automotive headlights. It is not so much indicative of the physical size as of operating voltage, which is typically 1.5 to 28 volts. Miniature lamps can be found in many industrial/commercial applications. They are used widely on indicator panels, circuit boards, automobiles, and elevators.

Identification of this family of lamps is generally by a two-to-four-digit number which may be either preceded, or followed by an alpha character. The identification process can be a nightmare as you most likely know. So when the numbers are not available on the lamp, do the following;

- Measure the length in inches or fraction of inches.
- Note the bulb shape; globe, tubular reflector etc.
- Determine the voltage:
    - A. directly from the lamp markings
    - B. using a volt meter at the socket of the lamp.
- Note the name and model number of the equipment.
- Be able to describe the application of the lamp.
- Find a knowledgeable miniature lamp distributor.

Figure 3-2 shows a sampling of commonly available miniature lamps.

FACTORS AFFECTING LIGHTING EFFICIENCY

Introduction

Dust, dirt, and grime, depreciated light output, lamp failures, and deteriorating lighting fixtures are at work on your lighting installation. All tend to steadily reduce the lighting efficiency from your lighting system.

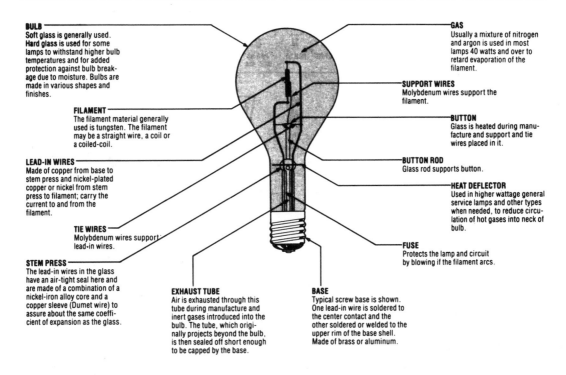

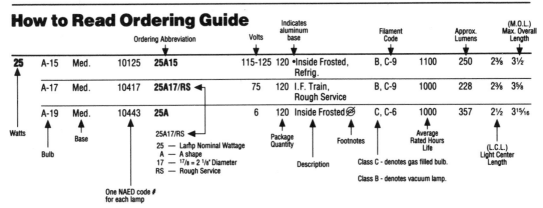

Courtesy of Osram Sylvania Lighting

Figure 3-1. How to Read a Lamp Ordering Guide

Dirt Depreciation

Dirt, dust, and grime settle on the walls, ceilings, and other light-reflecting surfaces. It may take weeks, months, or years, however, before the accumulation substantially reduces the amount of light that reaches the work area. And even though the amount of light is reduced, you continue to pay for total kilowatts of electrical power used.

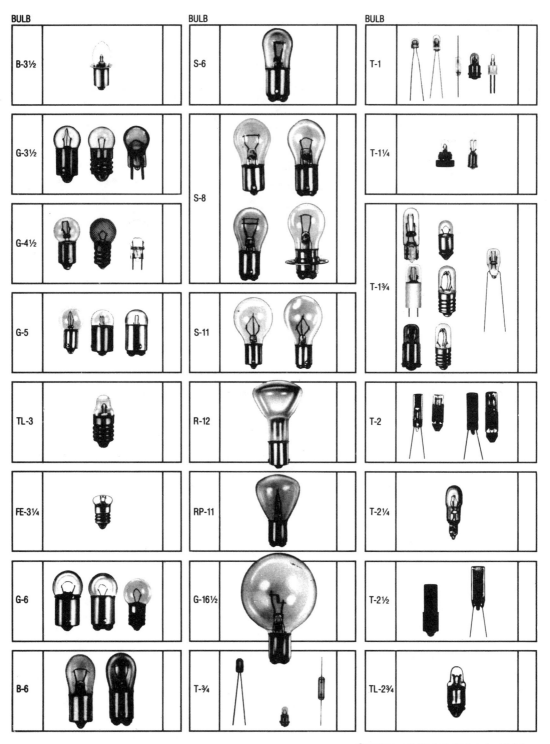

Figure 3-2. MINIATURE LAMP PRODUCT LOCATOR

Light loss due to dirt, dust, and grime depends on the type of lighting fixture used, the dirt conditions, and the time between cleaning. Losses range from 8 to 10 percent for a clean environment to 50 percent for severe industrial conditions.

Lamp Lumen Depreciation (LLD)

Nothing lasts forever. The more hours lamps burn, the less light they deliver. The Lumen Depreciation curve of every lamp is plotted by the manufacturer and provides light output at various points in the life of a lamp. A fluorescent lamp is typically producing only 80 percent of rated lumens at end of life.

Group Relamping

Group relamping is the changing of all lamps and the cleaning of all lighting fixtures in an area or entire facility.

Group relamping can be the most economical way of maintaining designed lighting levels, utilizing maintenance staff time, and purchasing of lamps.

Lamp life is measured in average rated life. The life in hours for a lamp represents the point when 50 percent of the lamps will have burned out. Group relamping at a point before average rated life avoids massive burnouts.

The following benefits will be realized by group relamping:

- A well lighted facility is a pleasant environment for all who use it.
- A maintained lighting system helps prevent accidents.
- Production will not be interrupted due to workers replacing lamps or inadequate light levels.
- Trouble calls for lamps out will be greatly reduced or eliminated allowing for better scheduling by the maintenance manager.
- Light levels will be at designed levels.
- Lamp costs will be reduced due to bulk purchases.
- Lamp storage and waste due to breakage will be reduced.
- The need for additional plug load lighting will be reduced.

Group relamping can be scheduled at a convenient time to utilize the entire maintenance staff. This can be done at plant shut down, summer holidays, extended breaks, or hours when the facility is vacant, usually in the evenings.

Top management may not see the benefits of group relamping because they perceive it as replacing a part that is still working. They must be educated to the benefits. Replacing a lamp with little life left before it burns out and cleaning the fixtures reduces costs.

Industry studies show the time required to group relamp and clean fixtures is 8-10 minutes per lamp. Spot replacing lamps, one at a time, requires 20-40 minutes per lamp. This substantial increase is required to locate lamps, ladder and cleaning supplies. These studies also show that when lamps are spot replaced fixture cleaning is rarely done. By not cleaning the fixture light levels will be reduced by the dirt depreciation of the fixture.

If you currently don't group relamp, use the following as a guide:

- 4000 hours of yearly operation: replace lamps every 4 years
- 6000 hours of yearly operation: replace lamps every 2.5 years
- 8000 hours of yearly operation: replace lamps every 2 years

The above assumes a 20,000 hour fluorescent lamp. For a dirty or dusty environment, Fixtures and or Lamps should be cleaned at 50% of rated life.

The Maintenance Manager can look at the lighting system as he does an automobile engine. Spark plugs and filters are replaced before they fail to prevent breakdowns and to maintain optimum system efficiency (MPG). Lamps can be viewed as the spark plugs and fixture cleaning as the car filters. By cleaning and replacing these items on a regular schedule, you will achieve optimum lighting efficiency (LPW).

INCANDESCENT LIGHTING SYSTEMS

A lighting system that uses incandescent lamps for anything but accent or display lighting should be replaced with another higher LPW system. Incandescent systems are only 18 LPW. Fluorescent or HID systems are in the 80-140 LPW range with lamp life in the 20,000 hours category. Incandescent life is in the 750 to 3500 hours range. Replacements for incandescent lamps and or fixtures is discussed in the energy management section Although usually trouble free problems can arise causing short life or reduced light output. Use the trouble-shooting guide to track down the problem.

Tungsten Halogen Systems

A Tungsten Halogen system is similar to an incandescent system in all respects. There are however two additional considerations for the maintenance staff to be aware of:

- Lamps must be handled only with the paper wrapper provided. If it is necessary to touch the lamp during installation, clean the lamp with alcohol and dry it with a clean, soft cloth before operation. Contamination of the quartz lamp by contact with human hands may result in reduced lamp life.
- This family of lamps operates under pressure and may shatter. Therefore all fixtures that use this type of lamp must come with a protective screen or other suitable enclosure. Do not operate this type of fixture without the provided protective equipment. Popular tungsten halogen lamps are called MR16's, Quartz, and low voltage lamps.

Incandescent and Tungsten Halogen Fixture Trouble-Shooting Guide

Problem	Possible Cause	Corrective Maintenance
Reduced Lamp life	Improper lamp and supply voltage	Change lamp voltage.
	Excessive vibration or shock	Use vibration lamp.
	Faulty lamp socket	Change socket.
	Voltage transients on the line	Remove transients.
	Fixture specs for max. lamp wattage	Use lower-wattage lamp.
	Moisture on lamp or in fixture	Cover fixture.
	Rated lamp life	Use longer-life lamp.
	Lamp burning time excessive	Use longer-life lamp.
	Excessive Heat in fixture	Use lower-wattage lamp.
	Improper lamp burning position	Use correct position.
	Improper fixture wiring	Rewire fixture.
Low Light Output	Proper lamp and supply voltage	Change lamp voltage.
	Faulty lamp socket	Change socket.
	Reduced lamp wattage	Use higher wattage lamp.
	Dirty fixture or lens	Clean lamp and fixture.
	Incorrect shape lamp	Change lamp shape.
	Improper fixture wiring	Rewire fixture.
Lamp Out	Lamp continuity	Change lamp.
	Faulty lamp socket	Change socket.
	Open fixture wire	Correct wiring.
	Open feed wire	Correct wiring.
	Open circuit breaker	Reset breaker.
	Switch off	Turn switch on.
	Malfunctioning occupancy sensor	Replace sensor.

TABLE 3-3. Incandescent and Tungsten Halogen Fixture Trouble-Shooting Guide

FLUORESCENT LIGHTING SYSTEMS

The fluorescent system is the most popular type of lighting in the industrial, commercial, and educational sectors. Commercially available in the late 1930s, this highly efficient and very adaptable form of lighting continues to evolve to meet today's lighting needs. A fluorescent system is somewhat more difficult to maintain due to the addition of a ballast contained in the lighting fixture.

Table 3-4, a Fluorescent Troubleshooting Guide shows some of the most common problems, their characteristics, possible causes, and the corrective action that should be taken. Because high voltages are common in fluorescent systems, it is recommended that only qualified personnel attempt to make electrical measurements or adjust wiring and that they use reasonable caution in doing so.

HIGH INTENSITY DISCHARGE SYSTEMS (HID)

The High Intensity Discharge family of lighting systems encompasses Mercury, Metal Arc, and High Pressure Sodium. Each type has different characteristics which must be under stood for proper maintenance.

Mercury Systems

The Mercury lamp has been around since 1902. It is the least efficient of the HID types at 60 LPW. Its long life of 24,000 hours is a maintenance asset. Due to its long life lamps tend not to be changed until they fail. This causes severe light loss due to lamp and dirt depreciation. In order to maintain adequate light levels group relamping of Mercury lamps should be considered at 16,000 hours of operation.

An example of mercury lamps long life is a case where a lamp burned for 45,000 hours. This may seem good but the light produced by the lamp was only 10 percent of its rating. This lamp was still consuming its rated wattage of 400 watts and producing the light of a 40-watt lamp. This is hardly an efficient use of energy.

When replacing mercury lamps the wattage, coating and base type must be noted. Replace with the same type to insure proper operation and color consistency. Mercury coatings come in clear (C), Brite White Deluxe (DX), and Warmtone (N) types. Wattage of lamp and ballast must be the same to ensure proper operation. Mixing lamp and ballast wattages will result in erratic operation and low light levels.

Mercury lamps in gyms must be in enclosed fixtures or be of the Safeline type. The enclosure will help to prevent objects striking and breaking the lamps outer jacket. If the outer jacket is broken, the lamp may continue to light causing

Problem	Possible Cause	Corrective Maintenance
Lamps Fail to Light	Blown fuse or open breaker	Reset and look for open circuit.
	Starter not reset	Reset or replace starter.
	Wrong Lamps	Check ballast label for correct lamp.
	Poor contact between lamp and socket	Adjust lamp for correct fit.
		Clean lamp sockets.
		Replace lamp sockets.
		Check alignment of sockets.
	Lack of cathode heat	Test for 2.5-4.5 volts at lamp socket.
	Normal end of life lamp	Replace with exact equal to preserve system appearance.
	Starter at end of life	Replace starter.
	Ballast not delivering minimum lamp requirements	Test or replace ballast.
	High humidity or dirt on lamps	Remove, clean, replace lamp.
		If condition prevails, enclose lamps.
Lamps Fail to Light	Extreme ambient temperature (hot or cold)	Correct ambient; if possible, change ballast and lamp for existing conditions.
	Low voltage at fixture	Test voltage at fixture; correct if possible; change ballast to existing voltage.
	Inadequate starting aid	Distance from lamp to fixture must not exceed one inch.
		Check grounding of fixture, ballast, and junction box.
	Improper wiring	Verify if wiring is correct according to ballast specification.
		Check for pinched wires at socket bar or loose connections.
	Ballast at end of life	Replace ballast with correct replacement.

Table 3-4. Fluorescent Troubleshooting Guide

73

Problem	Possible Cause	Corrective Maintenance
Slow or Erratic Starting	Wrong lamps	Check ballast label for correct lamp.
	Poor contact between lamp and socket	Adjust lamp for correct fit. Clean lamp sockets. Replace lamp sockets. Check alignment of sockets. Test for 2.5-4.5 volts at lamp socket.
	Lack of cathode heat	Test or replace ballast.
	Ballast not delivering minimum lamp requirements	Remove, clean, replace lamp.
	High humidity or dirt on lamps	If condition prevails, enclose lamps.
	Extreme ambient temperature (hot or cold)	Correct ambient, if possible. Change ballast and lamp for existing conditions.
	Low voltage at fixture	Test voltage at fixture; correct, if possible; change ballast to existing voltage.
	Inadequate starting aid	Distance from lamp to fixture must not exceed one inch. Check grounding of fixture, ballast, and junction box.
	Improper wiring	Verify if wiring is correct according to ballast specification. Check for pinched wires at socket bar or loose connections.

(continued)

Problem	Possible Cause	Corrective Maintenance
New Lamps Fail Within the First Few Days of Operation	Wrong lamps	Check ballast label for correct lamp.
	Wrong ballast	Verify that ballast ratings agree with existing conditions.
	Improper wiring	Verify if wiring is correct according to ballast specification. Check for pinched wires at socket bar or loose connections.
Short Lamp Life	Wrong lamps	Incorrect lamps will operate for a period of time with greatly reduced lamp life and performance.
	Short burning cycles	Life of fluorescent lamps is based on 3 hours per start. Shorter on time will reduce life.
	Poor contact between lamp and socket	Adjust lamp for correct fit. Clean lamp sockets. Replace lamp sockets. Check alignment of sockets.
	Lack of cathode heat	Test for 2.5 to 4.5 volts at the lamp socket.
	Ballast not delivering minimum lamp requirements	Test or replace ballast.
	Voltage at fixture to high or low	Limits for Normal Operation

Rated Voltage	O K Range
120	110-125
208	195-218
220	205-230
240	220-250
277	260-290
480	460-500

(continued)

Problem	Possible Cause	Corrective Maintenance
Short lamp life	Wrong ballast	Verify that ballast ratings agree with existing conditions.
	Wrong starter	Consult starter specs for proper type.
	Lamp operating in the glow state	One lamp in two-lamp circuit has failed. Lamp may be defective
	Improper wiring	Verify if wiring is correct according to ballast specification.
		Check for pinched wires at socket bar or loose connections.
Snaking, Blinking or Flickering	Impurities within lamp	Common for new lamp; should stop within first week of operation. If snaking persists, consult lamp manufacturer.
	Normal stroboscopic effect due to alternating current	Can be reduced by using Tri-Phosphors type lamps.
	Ballast not delivering minimum lamp requirements	Test or replace ballast.
	High humidity or dirt on lamps	Remove, clean, replace lamp. If condition prevails, enclose.
	Poor contact between lamp and socket	Adjust lamp for correct fit. Clean lamp sockets. Replace lamp sockets. Check alignment of sockets.
	Normal end of life lamp	Replace with exact equal to preserve system appearance.

(continued)

Problem	Possible Cause	Corrective Maintenance
Snaking, Blinking or Flickering	Low bulb wall temperature	Full wattage lamps require 50 degree F ambient temperature. Energy saving lamps require 60 degree F ambient temperature. Drafts will also cause this. Jacketed lamps may be required. Consult lamp manufacturer for specific lamp minimum temperatures due to the wide verity of types.
	Wrong ballast	Replace with correct ballast.
	Ballast not delivering minimum lamp requirements	Test or replace ballast.
	Voltage at fixture to high or low	Limits for Normal Operation

Rated Voltage   O K Range

Rated Voltage	O K Range
120	110-125
208	195-218
220	205-230
240	220-250
277	260-290
480	460-500

Problem	Possible Cause	Corrective Maintenance
	Improper wiring	Verify if wiring is correct according to ballast specification. Check for pinched wires at socket bar or loose connections.
Reduced Light Output	Normal maintenance	Light output is higher in the first 100 hours than published ratings. Lumen output then depreciates at published ratings.

(continued)

Problem	Possible Cause	Corrective Maintenance
Reduced Light Output	Dirty lamps, fixtures or room surfaces	Lamps, fixtures, reflectors and room dirt contribute to reduced light output. Changes in room surfaces. Remodeling or refinishing furniture will introduce new reflectance characteristics, which can cause light levels to change.
	Changing of ballast or lamp to other than originally installed	Lamps and ballasts vary on the amount of light they generate. Check lamp lumen output in manufacturer's catalogues. Check ballast for ballast factor which indicates percent of rated light delivered.
	Adding current limiting devices	Significantly reduce light output while reducing power consumed.
	Adding reflectors and removing lamps	Could significantly reduce light levels.
	Air currents	Lamps need to be protected from this. Cover lamps or use jacketed lamps.
	Ballasts not delivering minimum lamp requirements	Ballast must deliver rated lamp current; replace ballast if current is incorrect.
	Defective starter	If a lamp appears to hang in the glow state, the starter may be defective; replace.

(continued)

Problem	Possible Cause	Corrective Maintenance
Reduced Light Output	Voltage at fixture to high or low	Limits for Normal Operation
		Rated Voltage *O K Range*
		120 110-125
		208 195-218
		220 205-230
		240 220-250
		277 260-290
		480 460-500
	Improper wiring	Verify if wiring is correct according to ballast specification.
		Check for pinched wires at socket bar or loose connections.
Center of Lamp Dark Ends Lighted	Shorted starter condenser	Replace starter.
	Improper wiring	Verify if wiring is correct according to ballast specification.
		Check for pinched wires in entire fixture.
	Low bulb wall temperature	Correct ambient or enclose lamp.
Dark Areas or Spots on Lamp	Wrong lamps	Check ballast label for correct lamp.
	Poor contact between lamp and socket	Adjust lamp for correct fit.
		Clean lamp sockets.
		Replace lamp sockets.
		Check alignment of sockets.
	Lack of cathode heat	Test for 2.5 to 4.5 volts at lamp socket.
	Normal end of life lamp	Replace with exact equal to preserve system appearance.
	Normal lamp design	VHO 1500 ma T12 lamps are designed with pressure control chambers which appear as dark ends.

(continued)

Problem	Possible Cause	Corrective Maintenance
Dark Areas or Spots or Lamp	Normal attrition of cathode coating	As lamps age, the emissive coating is deposited at the end of the lamp creating dark or gray ends.
	Wrong ballast	Replace with correct ballast.
	Ballast not delivering lamp requirements	Ballast may have current crest factor higher than 1.7.
	Low ambient temperature	Under extreme cold lamps may exhibit this. Correct ambient or enclose lamps.
Difference in Lamp Colors	Normal maintenance	In addition to normal decrease in light output, a slight color shift can occur with age.
	Wrong lamp color	Check labels on lamps that appear different. Replace with correct color.
	Range of manufacturing tolerances	All processes require a range of tolerances. Due to the complex nature of phosphors, lamps can appear different. If this occurs, contact your lamp supplier so the lamps can be properly analyzed. Also, different lamp brands can be different colors of the same lamp type.
	Variations in fixtures	Variation in the surface or finish of the reflector or louver can cause color variation.
	Variations in temperature or air currents	Temperature has an effect on the color of a fluorescent lamp, as it does the output. If the color shift is serious, correct the ambient or enclose the lamp.

(continued)

Problem	Possible Cause	Corrective Maintenance
Difference in Lamp Colors	Voltage at fixture to high or low	Limits for Normal Operation

Rated Voltage	O K Range
120	110-125
208	195-218
220	205-230
240	220-250
277	260-290
480	460-500

	Improper wiring	Verify that wiring is correct according to ballast case.
Radio and Television Interference	Radiation from lamp to receiver antenna	Minimum distances from lamp to antenna receiver are 20 watts or less, 4 feet; 20-40 watts, 6 feet; 41-100 watts, 10 feet; over 100 watts, 20 feet.
Ballast Case Overheating	Wrong ballast	Replace with correct ballast.
	Ballast not delivering lamp requirements	Ballast may have current crest factor higher than 1.7.
	Poor contact between lamp and socket	Adjust lamp for correct fit. Clean lamp sockets. Replace lamp sockets. Check alignment of sockets.
	Normal end of life of lamp	Replace lamps.
	Shorted, sluggish or wrong starter	Replace starter.
	Voltage at fixture to high or low	Limits for Normal Operation

Rated Voltage	O K Range
120	110-125
208	195-218
220	205-230
240	220-250
277	260-290
480	460-500

(continued)

Problem	Possible Cause	Corrective Maintenance
Ballast Case Overheating	Improper wiring	Verify that wiring is correct according to ballast.
	Starter cycling	Replace starter.
	Short circuit	Replace ballast.
Ballast Noise	Normal ballast hum	Ballast used with fluorescent lamps produces a hum caused by the action of the iron core. How loud this hum will be depends on the sound rating of the ballast. Electronic Ballasts also hum due to the high frequency which they operate.
	Loose vibrating fixture components such as louvers lenses, and ballasts	Loose components vibrate, amplifying normal ballast hum. Tighten or replace as needed.
	Defective ballast	A severe hum is indicative of a defective ballast.
System Bothers Eyes or Causes Headaches	Psychological reactions	On some occasions complaints of fluorescent causing eye strain will be raised. The Journal of the American Medical Association stated "Fluorescent light is not harmful to vision. It should not cause eye strain if properly installed." Due to the variety of applications with fluorescent lamps a Lighting Consultant should be used to determine proper lighting. There are also laws concerning lighting and computer terminals that a consultant should be familiar with.

ultra violet light to be emitted. A Safeline lamp will extinguish itself if the outer jacket is broken avoiding ultra violet exposure to the environment.

Metal Arc Systems

Metal Arc lamps are the highest efficiency full-color light source. To properly maintain a lighting system with Metal Arc, lamps should be group relamped. This is to maintain designed light levels and color consistency. Group relamping is recommended at 70 percent of rated life. At that time a thorough cleaning of the system is also recommended to negate the effects of dirt depreciation.

When lamp replacement is necessary, care must be taken to install only the correct wattage, color, lamp orientation, and base type. Wattage must match the ballast as with all ballasted lamps. Lamp color is determined from the bulb finish. This can be clear, coated, or 3K.

Orientation is the burning position of the lamp determined by socket position. The common positions are; Base-Up, Base-Down, and Horizontal. A Universal type lamp can be lighted in any burning position. Lamps with specific burning restrictions (orientations) must be used only as specified. Failure to burn lamps in specified positions will cause short life, low light output, and severe color variations. Base types can be mogul or medium and are obvious choices. Care must be taken to replace any lenses that are incorporated into the fixture. The lenses are part of the UL listing which would be voided if missing.

Exceptions for lensed fixtures are base-up burning M400/U and M1000/U and the Pro Tech family of open fixture lamps which are UL approved without lenses.

Metal Arc lamps are highly efficient and should not be retrofitted.

High Pressure Sodium Systems (HPS)

The most efficient of the HID systems is HPS. It is known by its characteristic golden yellow color. Its color properties limits its use to none color critical areas.

Maintenance of this type system is the easiest. Lamp replacement must match the ballast. Base size must be determined to be medium or mogul. Lamp finish is clear or coated. No enclosures are required for this lamp type.

Group relamping is recommended for consistent lighting levels at 70 percent of rated life. Rated life is commonly 24,000 hours and can be increased to 40,000 hours with stand-by type HPS lamps. Due to its high efficiency retrofitting of a HPS system is not recommended.

Trouble-Shooting High Intensity Discharge Systems

To determine possible causes of problems with High Intensity Discharge lighting installations, a thorough analysis must be made of all operating conditions.

The best way to minimize difficulties is to follow a program of preventive maintenance. However, it is unreasonable to expect that preventive maintenance will eliminate all the causes of trouble which require corrective action. This makes it doubly important for the lighting engineer, plant engineer, or the lighting maintenance personnel to be familiar with the indications of trouble and know the methods to be used to correct the problems.

Table 3-5 is intended to serve as a guide in determining the possible cause or causes of the problem and to suggest corrective maintenance procedures. If a large percentage of lamps fail to operate in a new installation, it will generally be found that operating conditions are causing the trouble. In this case, the entire electrical installation should be checked thoroughly. Because high voltages are common in HID lighting systems, it is recommended that only qualified personnel attempt to make electrical measurements or take corrective measures and that they use reasonable caution in doing so.

Quick Troubleshooting Checklist for HID Systems

The following items should be checked before any replacement of lamps is made in an installation. Although many of the items may appear obvious or too simple a cause, the list should be checked over to avoid unnecessary changeouts and expense.

- Is power distribution system functioning properly? Is power switch actuated?
- Do circuit breakers remain closed or circuit otherwise activated to fixture when power is applied?
- Is electric eye or photocell OK?
- Is proper line voltage available at ballast primary? Is it within 10 percent?
- On multiphase circuits are all phases operating with all phase circuit breakers and ground functioning properly?
- Does voltage rating on ballast nameplate agree with line voltage and frequency available?
- Is ballast properly wired? How about wiring to capacitors and lamp socket?
- Is ballast properly grounded in fixture or to pole and mounting hardware?
- Is socket in good condition. Contacts shell and leads in good condition?
- Is lamp compatible with the ballast? Mercury on Mercury or Metal Arc, Metal Arc on Metal Arc, Lumalux on proper ANSI ballast, Metal Arc Swingline on single head CW or CWA mercury ballast, Unalux on mercury lag type ballast.
- Do lamp and ballast wattage ratings agree? Check nameplate and lamp etch.
- Does lamp operating position agree with specified position of the lamp?
- Are BU-HOR lamps being operated base up to horizontal?

HID Trouble Shooting Guide

Problem	Possible Cause	Corrective Maintenance
Lamp Will Not Start	Lamp loose in socket-improper insertion and seating	Inspect lamp base to see if there is any indication of arcing at the center contact button. Tighten lamp to seat it properly. If base is distorted and will not seat properly in socket, replace lamp.
	Electric eye inoperative	Replace electric eye.
	Defective or improper wiring	Examine wiring to make sure it agrees with wiring diagram on the ballast label. Check primary wiring to ballast and from ballast to socket to establish circuit continuity. Check connections to see that they are secure. Look for too small wire size, resulting in lowered voltage. Repair circuit.
	Voltage at fixture too low	Measure line voltage at input of ballast. For most types of ballasts, measured line voltage should be within 10% of nameplate rating. With many types of distribution systems, increasing loading or demand decreases available voltage at the ballast primary.

Table 3-5. HID Troubleshooting Guide

Problem	Possible Cause	Corrective Maintenance
Lamp Will Not Start	Improper ballasting	Proper ballasting is essential for dependable HID lamp operation. Any HID lamp will perform erratically or fail to start on an improper ballast. Make sure that the ballast nameplate data agrees with the line voltage and lamp used. Improper ballasting will generally cause a lamp to fail prematurely. Note: Mercury lamps will operate satisfactorily on ballasts designed for metal halide lamps of the same wattage.
	Defective ballast	A shorted ballast will generally cause the seals at the end of the arc tube to rupture with an indicative blackening in the seal area. A shorted condition may be due to shorted capacitors, shorted leads, or shorted windings.
	Improper lamp operating position (metal halide only)	The operating position should agree with the lamp etch. A BU-HOR lamp can be operated base up vertical to and including the horizontal, a BD only can be operated base down vertical to, approaching, but not including the horizontal. A lamp operated beyond the specified position may not start and may degrade lamp performance if it does start.

(continued)

Problem	Possible Cause	Corrective Maintenance
Lamp Will Not Start	Lamp has been operating: Cool down time insufficient (hot restrike)	HID lamps require a period of time to reestablish optimum starting conditions; after they have been operating and the supply line voltage has been interrupted. Mercury, sodium, and metal halide lamps require approximately one minute to cool before restriking.
	Improper ballast for lamp operating conditions	Under low temperature conditions, the ballast may not supply sufficient voltage to start the lamp. At -20 degrees F, for example, ANSI specifications state that 90 percent of lamps shall ignite when proper lamp voltages are impressed.
		The same problem may exist at a very high temperature. "Indoor" ballasts that are installed outdoors may not start the lamps. There isn't sufficient secondary voltage. Similarly "outdoor" ballasts are generally not adaptable for indoor use where the ambients are high. Check lamp environment against published performance characteristics.
	End of life ballast	The appearance of the ballast may give a clue to whether it is good or not. If charred, it may have been subjected to sustained excessive heat. Swollen capacitors indicate trouble. Check with appropriate continuity testers, ammeter, and voltmeter. Frequently, the failure process of a ballast is capacitor failure.

(continued)

Problem	Possible Cause	Corrective Maintenance
Lamp Will Not Start	Defective starter board	High-pressure sodium and compact Metal Arc lamps depend on an external starting circuit to provide high voltage pulses to start the lamp. If the pulse is not generated or is below specification, lamps will fail to start. A below-specification starter board may start a lamp initially but fail to start it a second time as the lamp's required starting voltage can increase for a short period while the lamp seasons.
	Mismatched starter board	Many ballast and starter board designs require that the two components be carefully matched to provide a starting pulse of the proper level. A low pulse will not start the lamp. A high pulse can damage the components of the ballast or the starting circuit.
	Defective lamps	Replace lamp. Note: Mercury and Metal Arc tube leakers can be identified by sparking the base with a Tesla coil. An arc tube will not ignite for a leaker.
	Major lamp defects	Sodium arc tube leakers can be identified by sparking the base with a Tesla coil. The arc tube will glow. The outer jacket will or may not glow. All other defects can be identified by visual inspection of the lamp.

(continued)

88

Problem	Possible Cause	Corrective Maintenance
Lamp Life Is Short	Lamp physically damaged. Outer jacket leaker.	Investigate the possibility of outer bulb damage from handling or transportation that may have cracked the glass. If air enters the outer bulb, the arc tube may continue to burn for 100 hours before failure. Check to see if the bulb is broken where the glass meets the base, due to twisting the lamp too firmly into the socket or scoring the glass where the socket inadvertently touches the bulb.
	Lamp physically damaged. Outer jacket leaker.	Look for broken arc tube or loose metal parts. Replace lamp. A leak in the outer bulb will cause oxidation of the metal parts inside. In high-pressure sodium, the dark getter material in the neck of the bulb near the base will turn white or disappear.
	Wrong ballast	Make sure that the ballast nameplate data agrees with the line voltage and the lamp used.

(continued)

Problem	Possible Cause	Corrective Maintenance
Lamps Flicker or Go Out, (This Can Be Intermittent Flickering or Cycling Where Lamp Starts Up, Then Goes Out)	Wrong ballast	With mercury lamps, improper ballasting can cause flickering or erratic operation. With metal halide lamps, the effect is generally noticed in start-up when the lamp ignites, starts to warm up, then extinguishes. This may be caused by improper voltage or current relationships delivered by the ballast. Wiring discontinuity can cause flicker. Under certain conditions new lamps may "cycle." Usually after three tries to start at 30-60 second intervals, lamps will stabilize and operate satisfactorily. High-pressure sodium lamps will cycle on and off if the ballast does not have sufficient open circuit voltage to sustain the lamp.
	High lamp operating voltage. Low open circuit ballast voltage.	Measure lamp operating voltage. Measure ballast open circuit voltage; replace as required.
	Variable voltage	Heavy motor loads or welding appliances on line can cause flickering during operation. Remove lighting circuits from the circuits servicing these devices. Provide voltage regulators. Check for loose connections. Use of constant wattage ballasts can frequently help this situation.

(continued)

90

Problem	Possible Cause	Corrective Maintenance
Lamps Flicker or Go Out	Hi-Spike lamp	Chemistry of a defective lamp sometimes causes the lamp to call for more voltage than the ballast can furnish and the lamp will start, extinguish, cool, and repeat the cycle. Replace the lamp.
	Hi-Spike ballast	Ballast may produce secondary reignition spike, which causes the lamp to cycle.
	HPS Cycler	As a high-pressure sodium lamp is burned for a long period of time, its operating voltage tends to increase. This voltage rise can reach a level where the ballast cannot sustain the lamp. When this point is reached, the lamp will exhibit a cycling on and off characteristic. This is normal end of life.
Lamp Starts Slowly (Arc Does Not Strike When First Turned On)	Hard starter	A hard starter is a lamp that will not start rapidly. It may glow for extended periods of time, destroying cathodes. It should be replaced after checking voltage and ballast.
Fuses Blow or Circuit Breakers Open on Lamp Start-Up	Overloaded circuit, high momentary transient current	Rewire to accommodate starting current of lamp/ballast combination.

(continued)

Problem	Possible Cause	Corrective Maintenance
Lamp Light Output Low	Normal light output depreciation throughout life	Refer to maintenance characteristics of the lamp in technical publications comparing light output versus burning time. If depreciation is within the published range, replace the lamp. If not, proceed to other areas of investigation.
	Incorrect ballast voltage	Check ballast nameplate to see if rating designation conforms to lamp rating description. Check line voltage at ballast; match ballast voltage range against input voltage. Check wiring connections for voltage loss points. Check socket contact point. Use CW ballast.
	Incorrect ballast output	Check ballast output to determine if it conforms to lamp requirements. If voltage and current do not stabilize in 5-10 minute warm-up time, ballast output is incorrect and replacement should made.
	Dirt accumulation	Check and clean lamp and fixture. Establish a maintenance program.
Arc Tube Becomes Black or Swollen over Lamp Life	Overwattage operation	Check for the possibility that the lamp is operated on ballast designed for a higher-wattage lamp. Overwattage operation can cause premature blackening.

(continued)

Problem	Possible Cause	Corrective Maintenance
Arc Tube Becomes Black or Swollen over Lamp Life	Improper ballasting	Check the ballast nameplate against lamp etch.
	Excessive current or voltage	Check voltage at ballast. Check for possibility of current or voltage surges which can damage arc tube, seals, or burn up connecting welds inside the outer bulb.
	Reflector problem	Reflector design may refocus radiant energy directly on the arc tube or other parts of the lamp, causing overheating. Fixture should be tested in laboratory.
	Glow state operation	Under certain lamp and/or ballast operating conditions, the lamp will go into partial discharge; check lamp/ballast.
Lamp Breakage Occurs	Scratched glass outer bulb	Investigate possible careless handling of fixture. If a pole type bulb changer is used, check the maximum diameter of the basket. If there are scratches on the bulb, make sure the socket does not contact the neck of the bulb, causing scratches in the glass.
	Improper insertion	Screw into the socket until firm contact is made. Then stop.

(continued)

Problem	Possible Cause	Corrective Maintenance
Difference in Lamp Color	Normal maintenance	In addition to a normal decrease in light output or brightness, color shift will occur as lamps age. Spot replacement of failures with new lamps may show very noticeable differences in lamp colors. A group replacement program minimizes this problem.
	Wrong lamp color	Check etch on lamps which appear different to see that they are actually the same colors. Replace with correct color lamp.
	Range of manufacturing tolerances	Any manufacturing process requires a range of tolerances in which to operate. With HID lamps, color differences may be caused by variations in quantities of materials in the arc tube. This is particularly true of metal halide lamps which depend on several metallic iodides in the arc stream to produce color. Color variations in phosphor coated types are also related to phosphor composition and thickness of phosphor layer. If variations are extreme, consult the supplier of the lamps.

(continued)

94

Problem	Possible Cause	Corrective Maintenance
Difference in Lamp Color	Range of manufacturing tolerances (continued)	Color is affected by wattage variations. Watts delivered to HID lamps may vary by + or − 7.5 percent according to ANSI standards. Interchanging lamps may minimize apparent color difference.
	Variations in luminaires	Variations in surface or finish of reflectors/lenses can introduce color differences. Dirty fixtures can create differences emphasizing the importance of adequate maintenance.
	Variations in environment	In common with fixture variations, color differences in ceilings, walls, floors, and furnishings, as well as other sources of illumination in the area can affect appearance of the lamp color.
	Super Metal Arc in incorrect socket	Horizontal super Metal Arc lamps have special Position Oriented Mogul Sockets (POMS) to assure proper arc tube orientation. If not correct, light output will be low and the lamp will fail early.

- Is proper voltage and current available at lamp socket as read on ballast tester? Do readings agree with lamp ballast specifications?
- Is adequate ballast voltage available for starting lamps at the required ambient temperature?
- Does lamp light on test ballast consistently and stay lit?

ENERGY CONSERVATION

Lighting loads in commercial and industrial buildings account for between 30-50 percent of the total monthly energy bill. This is due in part to long burning hours (in the order of ten to twenty hours per day) for a typical building.

Another factor is the quantity of the fixtures and their power consumed. By determining necessary footcandle levels and then measuring existing levels, an informed decision on reducing fixtures or components to meet the needs of the task performed can be made.

For most facilities burning hours and fixtures cannot be eliminated. For those situations a retrofit of existing fixtures or a redesign utilizing new fixtures will have to done. By careful selection of more efficient lamps, ballast and fixtures, lighting consumption can be reduced 25-50 percent. Some advanced techniques are described in detail and should be considered for all lighting retrofits. See Table 3-6 for IES Recommended Footcandle Levels.

Incandescent Lighting

Advances in incandescent lighting efficiency with halogen products with the trade name Capsylite can reduce existing lamp wattages by 50 percent. This line of lamps should be considered for all PAR or R lamps as well as standard A-Line lamp applications.

Compact fluorescent lamps are also worth investigating to replace incandescent lamps. These fluorescent lamps come in R and PAR enclosures and are suitable for many applications on a retrofit basis. When opting for compacts, footcandle readings should be taken to assure sufficient illumination after they are installed. Compacts can reduce energy consumption by 75 percent and last up to fifteen times longer than incandescent lamps.

Compacts are either electronically or magnetically ballasted. The magnetic units feature the classic blink, blink typical of starting of a preheat-type lamp. Electronic units start instantly. Most compact fluorescents do not dim and you should not attempt to dim them. There is a line of compacts that is designed to dim. These should be selected along with special dimming equipment for proper operation.

High wattage incandescent lamps (over 250 watts) should be viewed as a dinosaur and made extinct with an HID source. High pressure sodium should be used

Foot Candles*	General Activity (All Users)	Retail/ Merchandising	Healthcare	Industrial	Office/ Schools	Hotel/Motel and Restaurant
2 to 5	Public spaces with dark surroundings	• Parking lots • Entrances • Building perimeter	• Patient observation— night time • Parking garage	• Freight elevators • Railroad yard • General	• Parking lot • Parking garage	• Parking lots • Entrances • Outdoor canopy
5 to 10	Simple orientation for short temporary visits	• Storage—large items	• Corridors—night • Patient room— general day	• Storage— large items	• Janitorial supplies area	• Janitorial— general
10 to 20	Working spaces where visual tasks are only occasionally performed	• Storage—small items • Labels	• Corridors • Recovery room • Lobby • Reception • Task	• Storage—small items • Labels	• Storage • Corridors • Stairways • Elevators • Cafeterias	• Storage • Corridors • Stairways • Elevators • Intimate dining • Cashier
20 to 50	Visual oriented tasks with high contrast or large size	• Circulation • Dressing rooms • Stock rooms • Loading platform	• Dental suite— general • Examination room—general • Nurses station • Patient room reading	• Rough assembly and fabrication • Loading platform	• Circulation areas	• Dining • Bathrooms • Reading and work areas • Circulation areas
50 to 100	Visual oriented tasks with medium contrast or small size	• Check out areas • General merchandise • Circulation	• Nurses station— desk • Instrument tray • Local task • Critical examination • Ambulance port	• Medium assembly and fabrication	• Classrooms • Libraries • Science labs • Lobbies • Flags	• Kitchen • Front desk • Lobby • Cleaning crew
100 to 200	Visual oriented tasks with low contrast or very small size	• Show windows— general	• Operating room—general • Emergency outpatient • Tissue labs	• Difficult machine work • Assembly	• Lecture rooms • Shops • Music rooms	• Linen room— Sewing
200 to 500	Visual oriented tasks with low contrast and very small size over a prolonged period	• Feature displays	• Scrub areas • Delivery room— general • Autopsy table	• Fine assembly • Machine work • Inspection	• Graphic design department	
500 to 1000	Performance of very prolonged and exacting visual tasks	• Show windows— feature	• Surgical tasks	Drafting—high contrast (1000 F.C.)	Drafting—high contrast (1000 F.C.)	
1000 to 2500	Performance of very special visual tasks of extremely low contrast and small size		Operating table (2500 F.C.)	Drafting—low contrast (2000 F.C.)	Drafting—low contrast (2000 F.C.)	

*These values are summarized from the Illuminating Engineering Society of North America's published recommendations. The ranges have been compressed for brevity — for more specific values refer to the I.E.S. published recommendations.

Courtesy of Osram Sylvania Lighting.

Table 3-6. IES Recommended Illuminance Values

in warehouses and parking areas. Metal Arc should be used for interior, facade, and exit and egress areas.

As an example, a 300-watt incandescent produces 6000 lumens or 20LPW. An equivalent HPS lamp would be 70 watts and 6300 lumens or 90LPW. A Metal Arc lamp would be 70 watts and 5200 lumens or 74 LPW. HPS has an increased LPW and a longer life in comparison to Metal Arc. Metal Arc has a crisp white color. When retrofitting, use the above characteristics to guide your system selection. Changing to either HPS or Metal Arc will require a fixture change in most cases.

Exit Lights

A typical exit light will contain two 25 watt lamps. Burning 24 hours a day, 7 days a week they consume $43.5 in electricity costs. By retrofitting to compact fluorescent lamps, a savings of $30 per year per exit sign can be realized.

Additionally, the compact fluorescent lamps last 5 times longer than the incandescent lamp it replaces. Figure 3-3 shows how easy an exit sign retrofit can

The Twin Tube Exit Light Conversion Kit is a sub-assembly designed to field-convert 120V incandescent exit sign fixtures to compact fluorescent.

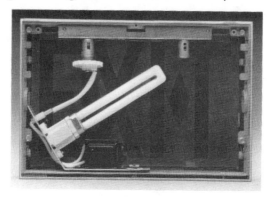

The Sylvania Twin Tube Lamp is highly efficient, delivering 50 lumens per watt. Rated average life is 10,000 hours, based on three hours of operation per start. Continuous operation, which is normal for exit signs, greatly increases lamp life. Finally, the compact fluorescent's virtual immunity to shock and vibration eliminates this type of premature lamp failure, a common problem with incandescent signs.

Energy savings are substantial. The Twin Tube fluorescent lamps use only 9 watts compared to 25 watts for incandescent lamps.

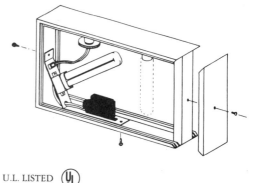

U.L. LISTED (UL)

Courtesy of Osram Sylvania Lighting.

Figure 3-3. Twin Tube Exit Light Conversion Kit

be. Just remove the lens, take out the existing lamps, and plug in the correct adaptor. This project can be done by any member of your staff.

Fluorescent Systems

Fluorescent lighting is a very efficient system, but improvements in its basic components have achieved dramatic efficiency improvements. Using T-8 bulbs with triphospor coatings coupled with electronic ballasts will reduce lighting loads up to 40 percent with no drop in light levels. The best part of achieving these savings is the ease of installation. Retrofitting to electronic ballasts requires a simple ballast change. Then changing the lamps to T-8 types completes the operation and the savings begin. An abundance of literature from manufacturers is available for these improvements. Fixture comparison charts indicate how a simple lamp and ballast change significantly reduces lighting consumption. Table 3-7, the fixture comparison guide, shows the characteristics for common fluorescent lamp and ballast combinations. Note that as the system watts are reduced, relative light output (RLO) is also reduced. The exception is for T-8 lamps and ballasts which produce the highest efficiency (RLO/watt) and the highest RLO. Table 3-8, the fixture conversion chart, will serve as an easy guide to the energy saving which can be achieved using the most efficient fluorescent equipment available. All the most common fluorescent fixtures are featured along with the saving that can be achieved with a simple ballast and lamp change.

Reflectors

Highly reflective mirror-like inserts can be installed in fluorescent fixtures to achieve a 50 percent power reduction per fixture. Typically, a reflector insert is installed in an existing fluorescent fixture and half the lamps are removed. This results in approximately 30 percent reduction in light levels. If a 30 percent reduction of lighting is within the IES guidelines, this method should be considered. Reflectors may also adversely affect light distribution causing uneven lighting and dark walls.

There are two types of reflectors commonly used: anodized aluminum with a reflectance of 85 percent, and silverized coatings that have a Reflectance of 95 percent. For comparison purposes a new fluorescent fixture has a reflectance of approximately 85 percent. These values decrease over time due to operating conditions. It is this decrease in reflectance caused by dirt, dust, and grime that is a major cause of decreased lighting levels. Group relamping and regular cleaning reverse light loss due to dirt and dust.

Due to the complex nature of a reflectorized system, a test of a reflector system is strongly recommended to determine if this technology is suitable for your facility.

FOUR-LAMP RECESSED

Troffer, Plastic Lens
77°F Test Room

LAMP TYPE	BALLAST	BALLAST FACTOR (1)	WATTS	RELATIVE LIGHT OUTPUT (RLO)(2)	RLO/W
SYLVANIA F40CW	STANDARD MAGNETIC	.95	174	100	100
SYLVANIA 34W/SS/CW	STANDARD MAGNETIC	.90	155	90	101
SYLVANIA 34W/SS/D841	STANDARD MAGNETIC	.90	155	95	107
SYLVANIA 32W/SSP/D841	STANDARD MAGNETIC	.90	144	93	113
SYLVANIA F40/CW	ENERGY SAVING MAG.	.95	162	101	108
SYLVANIA 34W/SS/CW	ENERGY SAVING MAG.	.88	139	89	110
SYLVANIA 34W/SS/D841	ENERGY SAVING MAG.	.88	139	93	116
SYLVANIA 32W/SSP/D841	ENERGY SAVING MAG.	.88	131	91	122
SYLVANIA 34W/SS/CW(3)	OPTIMIZER MAGNETIC	.95	116	87	130
SYLVANIA 34W/SS/D841(3)	OPTIMIZER MAGNETIC	.95	116	91	137
SYLVANIA 34W/SS/CW	ELECTRONIC	.75	119	86	125
SYLVANIA 34W/SS/D841	ELECTRONIC	.75	119	93	136
SYLVANIA FO32 OCTRON	T8 MAGNETIC	.95	132	104	137
SYLVANIA FO32 OCTRON(3)	T8 ELECTRONIC	.92	106	101	166

(1) Data in test normalized to ballast factors shown in this column for magnetic ballasts. Factors shown for electronic ballasts are measured values of sample.
(2) Relative light output based on initial (100 hour) rated lamp lumen output.
(3) Life reated at 15,000 hours. All other systems shown are rated at 20,000 hours.

Courtesy of Osram Sylvania Lighting

Table 3-7. Fixture Comparison Chart

Fixtures that are "Convertible" to the Octron System	Typical Wattages after Conversion to Octron System	Typical Pre-Conversion T12 System Wattages[1]	Estimated Change in Fixture Wattage with Conversion[2]	Estimated Change in Light Output with Conversion[3]
1-F20 (2' Fixture)	24 Watts w/Mag Ballast	32 Watts w/Std lamp	Down 25%	Up 7 – 10%
2-F20 (2' Fixture)	43 Watts w/Mag Ballast	50 Watts w/Std lamps	Down 14%	Up 7 – 10%
3-F20 (2' Fixture)	51 Watts w/Elec Ballasts (67 Watts w/Mag Ballast)	82 Watts w/Std lamps	Down 38%	Up 7 – 10%
4-F20 (2' Fixture)	57 Watts w/Elec Ballast (86 Watts w/Mag Ballasts)	100 Watts w/Std lamps	Down 43%	Up 7 – 10%
1-F30 (3' Fixture)	29 Watts w/Mag Ballast	43 Watts w/Std lamp 36 Watts w/SS lamp	Down 33% Down 20%	No Change Up 7 – 10%
2-F30 (3' Fixture)	48 Watts w/Elec Ballast (55 Watts w/Mag Ballast)	75 Watts w/Std lamps 61 Watts w/SS lamps	Down 36% Down 21%	No Change Up 7 – 10%
3-F30 (3' Fixture)	64 Watts w/Elec Ballast (84 Watts w/Mag Ballasts)	118 Watts w/Std lamps 97 Watts w/SS lamps	Down 46% Down 34%	No Change Up 7 – 10%
4-F30 (3' Fixture)	84 Watts w/Elec Ballast (110 Watts w/Mag Ballasts)	150 Watts w/Std lamps 122 Watts w/SS lamps	Down 44% Down 31%	No Change Up 7 – 10%
1-F40 (4' Fixture)	35 Watts w/Mag Ballast	55 Watts w/Std lamp 48 Watts w/SS lamp	Down 36% Down 27%	No Change Up 10 – 14%
2-F40 (4' Fixture)	62 Watts w/Elec Ballast (67 Watts w/Mag Ballast)	92 Watts w/Std lamps 78 Watts w/SS lamps	Down 33% Down 21%	No Change Up 10 – 14%
3-F40 (4' Fixture)	84 Watts w/Elec Ballast (102 Watts w/Mag Ballasts)	147 Watts w/Std lamps 126 Watts w/SS lamps	Down 43% Down 33%	No Change Up 10 – 14%
4-F40 (4' Fixture)	106 Watts w/Elec Ballast (133 Watts w/Mag Ballasts)	174 Watts w/Std lamps 156 Watts w/SS lamps	Down 39% Down 32%	No Change Up 10 – 14%
2-FB40 (2' x 2' Fixture)	60 Watts w/Elec Ballast (65 Watts w/Mag Ballast)	92 Watts w/Std lamps 78 Watts w/SS lamps	Down 35% Down 23%	No Change Up 10 – 14%

NOTES:
1. Estimate is based on fixtures with Standard Magnetic Ballasts
2. Compared to lowest wattage Octron system
3. Assumes same age of both old & new systems (4100K vs. CW). Does not consider immediate light level improvement provided by relamping.

Courtesy of Osram Sylvania Lighting

Table 3-8. Fixture Conversion Chart

Current Limiters

Current limiters are add-on devices to fluorescent fixtures to reduce light and power consumed. If a space is overlighted, this is a cost-effective method of reducing light and power. Using a Thriftmate device by Sylvania does not improve lighting efficiency. However, it reduces lamp wattage and light output. If Thriftmates are used, lighting loads and light levels will be reduced 50 percent.

Occupancy Sensors

The tendency for occupants to leave the lights on when exiting a space is an opportunity for conservation. Occupancy sensors can be installed in place of the manual light switch, in the ceiling, or on the wall. Then, when the room is unoccupied, the sensor turns off the lights, saving energy. Compared to manual switches, occupancy sensors typically save 15 to 30 percent of the connected lighting load.

Occupancy sensors are available in numerous packages to accommodate most applications. Standard adjustments include delay off times, and sensitivity to daylight and other background lighting.

Two technologies are presently available. Passive infrared that features moderate cost and acceptable motion-sensing properties, and ultra sound types which are higher in cost and offer superior sensing properties. Consultation with the manufactures of these devices will determine selection of the correct type.

Time Switches

A time switch is a very basic and SIMPLE device used to turn lights on and off at specific times. They are not widely used, however should be considered if appropriate. The time switch may be a clock system, computer (EMS) or other device that limits the operation of the lighting system. Savings of 15-50 percent can be realized with a time switch.

When evaluating time switches, don't overlook its effectiveness for exterior lighting. By limiting burning hours for facade, parking, canopy, and floodlighting utility bills will be trimmed. However you should evaluate the security factor when reducing or eliminating lighting in any area.

Photocells

A photocell is a device that uses ambient light to turn lights on or off. Insufficient light through the sensor closes the switch and the lights go on. When enough light is present, the switch opens and the lights go off.

Common applications are for street lighting and other exterior applications. When used in conjunction with a time switch, a photo cell will fine-tune the operation of exterior lighting by adjusting for overcast days and varying sunset times.

Dimmers

Dimming

Dimming a lighting system will reduce power consumed and light output conserving energy. Depending on the type of lighting system, the ratio of input power to light loss when dimmed will vary. Tables 3-9, 3-10, and 3-11 are the power and light outputs of the three lamp families when dimmed.

Incandescent Dimming

An incandescent system is very predictable when dimmed. Table 3-9 shows the changes in light output, power consumed, and life expectancy when an incandescent lamp is lighted at other than rated lamp voltage.

For example, a 100-watt lamp rated at 120 volts dimmed to 90% rated voltage will consume 85 watts, produce 72 percent of rated lumens, and last three times as long as a lamp at 100 percent of line voltage.

% RATED INPUT VOLTAGE	% POWER CONSUMED	% LIGHT	% LIFE
120	125	125	50
100	100	100	100
90	85	72	300
80	70	50	1500
75	60	40	2000

Table 3-9. Incandescent Dimming Effecting Power, Light Output, and Life

Fluorescent Dimming

A fluorescent system is a complex one to dim. Special dimming equipment must be installed to insure maintained lamp life. See Table 3-10 for fluorescent dimming characteristics.

For example, below 50 percent fluorescent dimming systems tend to flicker and may reduce lamp life. If this approach is attempted, verify that all components of the system, lamp ballast, and dimmer are comparable. Recent advancements in fluorescent dimming technology allow this family of lamps to be dimmed to 2 percent. Equipment to achieve these low levels is highly specialized. Consultation with a lighting professional is recommended to install a system of this type.

% POWER CONSUMED	% LIGHT	% LIFE
100	100	100
80	80	100
75	75	100
50	50	100

Table 3-10. Fluorescent Dimming Power, Light Output, and Life

HID Dimming

Due to the nature of HID sources, dimming is not usually recommended. However, recent advances in ballast technology and lamp research have produced reasonable results with HID dimming.

Table 3-11 describes the characteristics of dimming HID sources.

Dimming of any of the discharge systems (fluorescent, HID) requires very careful selection of all components. Consultation with all manufacturers is strongly recommended for this conservation method.

Daylight Controls

Daylight controls uses a dimming ballast to maintain constant light levels in a space which has a daylight component. Dimming the electric lighting system when adequate daylighting is present can reduce power consumed by 30-70 percent. The sensor is adjusted for the desired level and then automatically maintains that level through dimming or full power.

Utility Rebate Incentives

Due to recent rulings by the Public Service Commission (PSC), public utilities are being asked to reduce electric power generation and still meet rising electric power demands. Through conservation efforts, these goals are being met. Techniques including rate recovery, least cost planning, DSM programs and direct incentives are used to meet the PSC demands. To encourage efficient equipment usage, utilities are paying incentives to customers who install this lighting equipment. The lighting rebates are usually based on a menu of acceptable technologies and usually include: electronic ballasts, occupancy sensors, compact fluorescent lamps, and new efficient lighting fixtures. Custom incentives may also be available based on the total number of watts saved. Any lighting retrofit should not be attempted until a call is made to the local utility to inquire about an incentive plan for lighting.

LAMP	COLOR SHIFT	DEGREE OF COLOR SHIFT	%POWER	%LIGHT
MERCURY	NOT DISCERNIBLE	NEGLIGIBLE	75	70
			50	40
METAL HALIDE	BLUE	DRASTIC	75	64
			50	27
HIGH-PRESSURE SODIUM	ORANGE	NEGLIGIBLE	75	70
			50	40

Table 3-11. HID Dimming Power, Light Output, and Color Shift

Case Studies in Lighting Retrofits

Case I

A computer company in Massachusetts decided to update their old lighting system. The existing system was experiencing lamp failures of 28 percent and ballast failure rate of 16 percent over a three-year period. By upgrading the system to T-8 lamps this company was able to modernize its lighting and take full advantage of the utility rebates available.

The old three lamp system consumed 138 watts compared to 78 watts when the system was updated to T8 lamps and electronic ballasts. Similarly, the four lamp fixture consumption was reduced from 181 watts to only 108 watts. This translated into an energy savings of $285,000 over the life of the lamps which is 20,000 hours. Payback for this retrofit was 1.75 years with the utility rebates.

Case II

A local school board in Michigan decided to renovate the lighting in the school district. After evaluating several proposals they decided to use the T8 type lighting. In addition to a reduction of 30 percent in energy costs, footcandle levels were maintained at 75. This footcandle level was the same as the original lighting system. The superintendent and his staff at this school system was "extremely pleased" with the new system. Installed were 3080 FO32/41K T8 lamps.

Energy costs saved were $118,000 over the 20,000 hours of lamp life. This translated into a return on investment (ROI) of 47 percent.

Case III

A hospital complex in upstate New York retrofitted their lighting system in order to keep the rising costs of health care to a minimum. By installing 40,000 FO32/41K T8 lamps and 1,000 exit light kits they significantly reduced their operating costs. The local utility wrote a rebate check for over $325,000 and the hospital reduced its electric bill by over $36,000 per month. The hospital was so pleased with this project it plans to do selective retrofitting in 15 more of their of their buildings. Payback for the project was 1.1 years.

Case IV

A well known hotel and casino in Las Vegas, Nevada has over one million square feet of public area. Included are 50,000 sq/ft of casino space 250,000 sq/ft of convention meeting rooms and 750,000 sq/ft of guest rooms, restaurants, and shops. The director of maintenance of the hotel set out to determine the scope of the lighting load after reading about utility incentives available.

He discovered that the convention facility used 1000 300-watt flood lamps and 1500 F40/CW lamps. By replacing the existing 300-watt lamps with 150-watt

Capsylite lamps and the F40's with F40/D35/SS an energy saving of $45,000 per year was realized. The fluorescent lighting in the general work area presented another opportunity to cut costs. A survey showed the hotel had more than 6,000 F40/CW lamps operated 24 hours a day. By switching to a FO32/D35/SSP lamp another 10.5KW was saved.

The hotel's 2,200 guest rooms were an area of major concern. It was extremely important that these rooms have adequate light levels yet, lamps must be operational at all times. The solution was to replace three 100-watt A lamps with a 72 MB halogen lamp. This replacement reduced lighting in the guest rooms by 28 percent for an annual saving of $15,000 in this area. The result of this carefully planned lighting replacement resulted in an annual saving of $85,000 annually for this hotel. Payback was eight months with no utility rebates.

Computer Models for Lighting Energy Savings

A complete system analysis must be done to effectively evaluate any proposed technology. Figure 3-4 is an example of a detailed study available from a computer analysis. Here the vital information is indicated and contains the following:

- Lamp yearly operating hours
- Cost of power in kwh
- Demand charge if any
- Air conditioning saving
- Increased heating expense
- Utility rebate, if any
- Annual return
- Total system cost including installation

The computer model can then be evaluated by the maintenance manager to determine the best retrofit for his facility.

Lighting has come to the forefront as the most cost effective method to reduce our nations energy consumption. Utilities have found that lighting is the least capital intensive, easiest to implement and has the fastest payback of any energy efficient technology. Changing a lamp and a ballast can achieve 40 percent gain in efficiency. An exit light conversion from incandescent to fluorescent reduces consumption by 70 percent. Changing a 150-watt PAR lamp to a 75 watt halogen Capsylite PAR lamp results in an immediate 50% reduction in power with virtually no light loss.

Green Light Programs

The Green Lights program sponsored by the U.S. Environmental Protection Agency (EPA) has enlisted companies to audit their lighting, install efficient equip-

System Data	Present	Proposal 1	Proposal 2
Lamp Description:	F40CW	F40CW/SS	FO32/31K
Ballast Type:	Standard	Standard	Electronic
Number of Lamps:	10,000	10,000	10,000
Annual Savings - Lighting			
Total Present System Lighting Load (KW)		460.00	460.00
Total New System Lighting Load (KW)		390.00	265.00
Load Reduction Due To Lighting (KW)		70.00	195.00
Load Reduction For 4,000 Hrs/Yr		280,000	780,000
Savings at $ .080 per KWH		$ 22,400.00	$ 62,400.00
Demand Charge Savings at $10.000 /KW		$ 8,400.00	$ 23,400.00
Annual Savings - Air Conditioning			
Reduced A/C Load per Year (Therms)		6,211.66	17,303.91
Reduced A/C Capacity Required (Tons)		19.91	55.46
Savings on Reduced A/C Load		$4,141.28	$11,535.68
Annual Expense Heating			
Heat Load Reduction (BTU/Hr)		238,910	665,535
Total Heat Load Reduction per Yr (Therms)		9,556.40	26,621.40
Add'l Fuel Req./Yr of No. 4 Fuel Oil		2,636	7,342
Annual Cost for Add'l Fuel		$210.88	$587.36
Annual Savings Summary			
Total Lighting Energy Savings		$30,800.00	$85,800.00
Plus Annual Savings in A/C Costs		$4,141.28	$11,535.68
Less Annual Add'l Fuel Cost		$210.88	$587.36
Total Annual Energy Savings		$34,730.40	$96,748.32
Return On Investment			
Total Energy Savings Over Life		$173,652.00	$483,741.60
Less Total Investment for New Lamps/Sys		$25,000.00	$150,000.00
Plus Utility Rebate		$10,000.00	$50,000.00
Net Return on Investment		$158,652.00	$383,741.60
Annual Net Return		$31,730.40	$76,748.32
Return On Investment (%)		211.5	76.7
Pay Back Period of Proposal (Months)		5	12
Emissions Reductions *– Carbon Dioxide		224.00	624.00
tons per year ––+ Sulphur Dioxide		1.63	4.55
(avg. all sources) *– Nitrogen Oxide		0.86	2.40

Courtesy of Osram Sylvania Lighting.

Figure 3-4. Computerized Model to Evaluate Retrofits Using the Sylvia Software

ment and save the environment by reducing lighting energy consumption hence, the name Green Lights. (To learn more about how your company can become a Green Lights Partner, call the EPA at (202) 775-6655 or fax them at (202) 775-6680.)

LEGISLATION

Significant revisions in federal, state, and local laws are being made as of this writing. To ensure compliance with current ordinances, you should check with the local building department before undertaking any major retrofit.

Local Conservation Laws

Most municipalities have code requirements for minimum footcandle levels. These are usually specified for interiors of schools, office, exit lighting, exterior and emergency lighting. These codes should be consulted before modifying any of these areas to insure code compliance.

State Conservation Laws

Most states have existing energy codes for lighting loads. The most common lighting code is the IES/ASHRAE 90.1. This document outlines maximum lighting loads based on the task being performed in the space and is expressed in watts per square foot (W/SF). Two W/SF are typical for an office building.

Other prominent state codes include California Title 24 and New York's Component Lighting Law. These codes differ greatly in approach in achieving efficient lighting. Due to the variety of methods in lighting codes your respective state agency should be contacted to confirm code compliance.

Federal Conservation and Ballast Laws

Since 1987, the manufacture and installation of "standard" magnetic ballasts has been banned in the United States. This was done under the National Appliance Energy Conservation Act of 1987 (NAEC). Since then, only energy-efficient fluorescent ballasts can be manufactured. Electronic types of energy efficient ballasts are considered the most efficient.

A comprehensive energy bill called the National Energy Efficiency Act (NEEA) passed in 1992. This legislation bans the manufacture of inefficient fluorescent and incandescent lamps. For example, the most popular type of fluorescent used today is the F40/CW lamp. This bill outlaws the manufacturer of this product in January of 1996. Another very popular lamp the F96/CW will be outlawed in June of 1994. Both of these lamps will have to be replaced by their energy saving equivalent. Lamps which can be used are the F40/CW/SS and the F96/CW/SS. Incandescent PAR and reflector (R) lamps will also be restricted.

For example, a 150 PAR/FL or Spot will be banned in January, 1996. The only lamp available will be a Halogen PAR Capsylite spot or flood type. A complete listing of the affected lamps is on Tables 3-12, 3-13 and 3-14. Table 3-12 depicts the fluorescent lamps specifications in the NEEA. Fluorescent lamps which meet these specifications will continue to be manufactured. Lamps that do not, can no longer be made after the effective date.

Lamp Type	Lamp Wattage	Minimum CRI	Minimum LPW
4-foot	> 35 watts	69	75
	< 35 watts	45	75
2-foot "U"	> 35 watts	69	68
	< 35 watts	45	64
8-foot Slimline	> 65 watts	69	80
	< 65 watts	45	80
8-foot High Output	> 100 watts	69	80
	< 100 watts	45	80

The following types of lamp are excluded from the above regulation:

- CRI of 82 or greater
- Lamps for plant growth
- Cold temperature types
- Reflectorized/aperature
- Impact resistant
- Reprographic service
- Colored
- Ultra-violet

Table 3-12. Federal Efficiency Standards for Fluorescent Lamps

Table 3-13 outlines the labeling required for all regular incandescent (non-reflector) lamps.

Table 3-14 shows the requirements of reflector incandescent lamps. Here again reflector lamps (R PAR) which fail these specifications can no longer be manufactured after the effective date.

Medium-based incandescent lamps will be required to display the following clearly printed on each package:

- Voltage
- Life
- Lumens per Watt (LPW) based on 120 volt operation

Table 3-13. Federal Efficiency Standards for Regular
Incandescent Lamps (Nonreflector)

Reflectorized Lamp Wattage	Minimum LPW
40-50	10.5
51-66	11.0
67-85	12.5
86-115	14.0
116-155	14.5
156-205	15.0

Lamps excluded: Minature, Decorative, Traffic signal, Marine, Mine, Theatre, Railway, Colored, Par20, R20 and smaller, ER, and BR shapes.

Table 3-14. Federal Efficiency Standards for Reflectorized Incandescent Lamps

The National Energy Efficiency Act will save business billions of dollars annually, as estimated by the US. Department of energy.

State and Federal Lamp Disposal Laws

Currently there are no federal laws regulating the disposal of lamps. There is a concern however about the small amount of mercury contained in fluorescent tubes and mercury lamps. These two families of lamps contain mercury to produce light. After the useful life of these lamps the mercury is still contained inside. Testing of a crushed lamp shows that mercury levels are well below government recommended levels. Therefore no special requirements are needed for these lamps as of this writing.

However, the state of California requires that disposal of more than 25 fluorescent or mercury lamps be treated as low-level hazardous waste. As of this writing, no other state has a similar requirement.

Federal and State Ballast Disposal Laws

Fluorescent lamp ballasts produced before 1976 most likely contain a PCB-filled capacitor. PCBs are a toxic substance banned in 1976 under the Toxic Substance Control Act of 1976 (TSCA). This Act specifically allows disposal of a PCB-containing capacitor in fluorescent transformers (ballasts) in a sanitary landfill. TSCA also specifies that a leaking PCB-type ballast must be disposed of in a specific manner and is classified as hazardous waste.

Conversely, the Superfund Act of 1980 limits disposal of PCBs to 16 ounces in a nonhazardous landfill. This limit of 16 oz. is the equivalent of 16 fluorescent lamp ballasts containing PCB capacitors.

At this time, only in the state of California is there a definite answer. California law specifically bans more than 16 fluorescent ballasts containing PCBs in a sanitary landfill. As of this writing, no other state has ventured into this grey area. Maintenance managers must stay abreast of all new laws and local requirements regarding PCB disposal.

4

Energy Conservation and Intelligent Buildings

John A. Bernaden

Building intelligence starts with a monitoring and control computer network called the building automation system (BAS). Operating with temperature, pressure, humidity, time, and other data, this system is designed to automatically maximize the comfort of the occupants, while minimizing the energy used by the building's heating, ventilating, and air conditioning (HVAC) equipment. This system can also be used to achieve economical, efficient lighting control. Other system functions may include automated security and fire protection, depending on what type of integration of these functions is permitted under local codes. In addition, computerized building maintenance management can be included as an integral part of BAS operations so it uses actual equipment runtime totals rather than calendar dates to schedule maintenance.

Most buildings today have the beginnings of a BAS in the form of thermostats, time clocks, smoke detectors, door alarms, and other devices. But these are very limited in scope compared to the total environmental control provided by a sophisticated BAS with distributed. processing on a local area network bringing the whole building together. Installation of a building automation system is therefore an essential first step in energy conservation management for every intelligent building. If you can't measure it, you can't manage it. However, as is often true in life, the simplest element sometimes has the greatest impact on what we do. The same is true for energy conservation.

SETTING AND CONTROLLING BUILDING THERMOSTATS

Maintaining the energy conservation system in any building, particularly an intelligent one, begins at the thermostat. Here most of the complaints begin: "too hot," "too cold," "not enough air!" So you should start here to determine the solutions.

Because the room thermostat is visible and easily accessible, people sometimes tamper with it. It is one of the few outlets for their frustration about an uncomfortable environment. Tampering may range from putting an ice bag on it in winter, or a cigarette lighter under it in summer, to actual tampering under the cover.

When someone calls to complain about the building temperature, modern building automation systems allow you to conveniently display and adjust any room temperature on a computer screen in seconds—without ever leaving the office. But if you continue to get complaints from the same building zone, an old-fashioned personal inspection is in order.

After ruling out tampering, check for proper location of the room thermostat. Be sure that it does not sense any drafts from doors or loose windows, or artificial heating effects of office equipment like copiers, FAX machines and computers. It would also be prudent to observe the occupants' clothing, especially during the changeover season. When changing heating to cooling or vice versa, adjusting room temperatures to match changing clothing attires can be a challenge.

A summer space temperature of 78°F and a winter space temperature of 68°F is still encouraged in the ASHRAE comfort standards. During each season, these higher and lower setpoints are judged to be reasonably acceptable to occupants as well as energy efficient.

Humidity also affects occupant comfort. Facts about bacteria growth and indoor air quality have made it mandatory to reduce the extremes. Relative humidities of 30 to 60 percent are now being accepted, recommended and even required by code. The conditioning systems in many existing buildings were designed to produce 50 percent RH in the spaces on a year-round basis. It has, therefore, become good energy conservation practice to use no "new energy" to maintain summer relative humidities below 60 percent or winter relative humidities above 30 percent. Practically, this means do not add moisture in the winter to keep the space RH above 30 percent. Do not maintain supply air temperature below that required to handle the sensible cooling for the purpose of extracting moisture unless space relative humidities exceed the 60 percent limit.

However, considering the complex dynamics of buildings today, adhering to simple rules of thumb for temperature or humidity may unknowingly be wasting energy. For example, before deciding on a space temperature, question the following:

- Does the system really use energy to maintain the temperature above 68°F during the heating season? Consider interior zones separately because they have no heating season. Temperatures above 68°F may be economical there year-round.

- Can a higher interior temperature help the exterior by simple interior-to-exterior heat transfer? Consider up to a 5°F temperature differential to convect excess interior heat toward perimeter zones without affecting occupant comfort.

Change the room thermostat settings to 68°F heating, 78° cooling, ONLY IF IT SAVES ENERGY. For example, during the spring changeover season, assume conditions are 63°F outside dry bulb, RH 30-100 percent with interior zone cooling required. In most cases, 63°F air delivered to these spaces will handle the cooling load; therefore, do not energize mechanical cooling. Thermostats set at 68° will use all outdoor air to maintain space temperatures. Switch the system to 78°F and turn on cooling at "first complaint." This is a subtle way to condition occupants to accept the extreme difference between the heating and cooling season. More important, it avoids using heating energy to drive temperatures up to 78°F.

Methods of Seasonal Changeover in Temperature

Complaints concerning seasonal changeover create one of the biggest obstacles to many maintenance managers using this fundamental energy conservation technique. Computerized control with modern building automation systems and digital thermostats have helped alleviate some of these concerns. Previously, changeover of all thermostat set points had to be done manually by maintenance personnel, because most thermostats have closed adjustments to maintain the integrity of energy conserving settings. It also involved making a very educated guess as to when the change should be made. Many authorities recommended specifying a firm cooling season, typically from May 15 through September 15. But once the change is made, occupants must live with a potentially uncomfortable setting until the next changeover time. Central changeover can be accomplished from one location with direct digital temperature control, permitting more changeovers per year, if necessary. In addition, most building automation system computer programs take some of the guesswork out of the changeover process, in addition to smoothing out the transition.

Since pneumatics still control almost 75 percent of temperature systems, that method of changeover deserves discussion, too. Avoid using an outside air transmitter for changeover. During the intermediate season, this would force the heating systems on and off possibly several times a day. Use dual temperature thermostats; their set points can be altered by changing the pneumatic supply air pressure. Submaster thermostats allow temperature transitions (68°F to 78°F) to be made gradually. This will reduce complaints and may save energy.

However, master-submaster pneumatic systems are a difficult control system conversion. Supply pressure changes, used to cause dual thermostats to change set points, may upset other existing single temperature controllers. Separate the pneumatic supply air circuits for the central system and room controls. If two-position changeover is used to change from summer to winter settings, execute the changeover during unoccupied periods. All proportional room thermostats have a throttling range of approximately $+/- 1°F$ or $2°F$. With the savings that can be realized for each degree change, the thermostat should be set to drift in the direction of energy savings (up for summer, down for winter).

Another possible cause of the initial hot or cold call in the beginning of this scenario could have been failure of the thermostat itself. As a pneumatic temperature control system ages, some maintenance managers expect that thermostats should normally drift from their setpoint. They accept the need for recalibration with increasing frequency as standard operating procedure. However, recent studies by ASHRAE and controls companies show that given a "perfectly" clean air supply, a thermostat will almost never need recalibration. The definition of "perfectly" clean is what has changed.

Common contaminants in the air stream such as oil, moisture, and dust typically cause system failure. The use of better air compressors and more extensive filtration systems cannot be emphasized enough. For example, previously filters which removed oil vapors were thought to be sufficient. Now we know that very small atomized oil particles easily pass through such vapor filters and can significantly affect a control system. Special costly charcoal filters which require more frequent replacement eliminate atomized oil particles. Because it is less expensive than constant service calls to calibrate thermostats, install a charcoal filter after all other filters in the air line. And consider using the best low-oil transmission air compressor for system-wide energy savings and reduced maintenance.

At the same time, clean contaminated air lines and pneumatic controls. In the past, CFC refrigerants were bubbled through the system to clean them. Obviously, that is no longer environmentally advisable. Replacement HCFC refrigerants may work equally well, although the environmental soundness of bubbling these chemicals through an air system and into the atmosphere is yet to be determined, too. Depending on the age and degree of contamination, cleaning may not completely restore the air delivery system and controls. For example, those particles of atomized oil stick to the walls of an air line, dry out; and then another layer builds up atop of it. This condition, like hardening of human arteries, slowly restricts the air supply. This eventually causes thermostats to seemingly drift out of calibration. What is changing isn't the thermostat adjustment, but actually the amount of air supplied.

Electric and electronic thermostat sensors typically fail completely one day, rather than slowly drifting out of calibration. However, modern computerized control systems feature a program that automatically "tunes" or calibrates a control loop. This not only eliminates the timely process of calibrating and tuning numerous control loops during building start-up, it also eliminates that servicing task as well, if necessary.

One final common sense measure in regard to saving energy at the thermostat on the wall: Remove all visual set points and thermometers so the occupant is not aware of existing temperatures. The feeling of comfort is not simply physical. Studies show that providing the most comfortable environment for people can improve their productivity. As perception of comfort is partly psychological, the lack of visual set points allows more people to assume they're comfortable, and thus renders them more productive.

Determining the Breakeven Temperature

The breakeven temperature is the outdoor air temperature that produces a transmission infiltration loss through the perimeter wall that is equal to the heat gain in the perimeter area from people, lights, and so on. Using 15 to 20 feet of the external section of the building to generate internal heat gain, and figuring standard people and lighting loads, the breakeven temperature calculates to be in the mid-40s for buildings with an overall "U" factor of .8, to below zero for buildings with "U" factors below .25. Some buildings surveyed during below zero (−0) weather were actually requiring cooling when the building was occupied and the lights were on. High-tech office equipment also compounds these internal heat loads beyond original design considerations.

With an average breakeven temperature in the mid-40s, we can assume that heat must be added to the building only during those occupied hours below about 45°F outdoors. The rest of the heating energy is used during the unoccupied cycle which includes the warm-up period. Let's assume that 75 percent of the total heating energy is used during the unoccupied cycle, 25 percent during the occupied cycle. For lower breakeven temperatures, this 25 percent may be reduced appreciably in actual savings per degree of inside temperature setback.

Added to these savings, a lower breakeven temperature lessens the number of hours heating will be required over a season. In northern climates, these savings often double the savings due to transmission losses below breakeven. The total savings, however, are usually less than 1 percent of the total heating bill for each 1° decrease in space temperature.

These savings can be unwittingly reduced, however, if the system is not switched over to the higher space temperature as soon as the building internal heat gain equals the transmission losses. There are two reasons for this:

- More cooling energy will have to be used at temperatures above "breakeven" to maintain the lower temperature. This could work in your favor if it's a reheat system, but could work against you if it's a variable air volume system.
- Using higher space temperatures has the mass of the structure warmed up to a higher temperature at the beginning of the unoccupied cycle (more heat stored). This stored energy will be given up to the space and keep the night or unoccupied heating device (fan system, etc.) off for a longer period of time.

As you can see, there is no simple answer. Under certain climatic conditions and building configurations, it's possible that a temperature reduction during the occupied cycle could waste energy. The only fact known for sure is that you must consider reducing the temperature during the "heating season" which is now defined as those hours below the "breakeven" temperature.

Building type and construction will, of course, vary these percentages. For instance, a "greenhouse" (glass and metal curtain wall) with excessive leakage would have a higher breakeven temperature, less thermal storage, lower "U" factor, and therefore, a greater percent of total heating energy used during the occupied period, but not enough to raise the savings much over 1 percent per 1°F.

Using Outdoor Air

ASHRAE minimum outside air requirements swung dramatically during the last decades, first for energy conservation reasons, and then indoor air quality crises. Obviously, using less "fresh" outdoor air means using less energy consumption heating or cooling it (depending on the season). The previous standard minimum stated that the outdoor air quantity should not be less than 5 CFM per person. But some systems tightly designed to this minimum literally starved occupants for proper oxygen supply. When slightly out of balance, these systems have become a maintenance manager's nightmare.

For example, assuming a design of 5 CFM/person, with average sq. ft./person at 125, and an air handling system supplying 1.0 CFM/sq. ft.

$$\frac{5 \text{ CFM/person}}{125 \text{ sq. ft./person} \times 1.0 \text{ CFM/sq. ft.}} = 4\%$$

Four percent is usually so little outside air that toilets and kitchen exhausts will move more than 4 percent out of the building, causing negative pressures. Today, ASHRAE standards require 20 CFM/person, so about 15 percent of air moved by the system should be outdoor air.

To maintain the best trade-off of energy conservation and greater outdoor air intake, you should carefully review your system design. Preheat coils are located in the outdoor air section of a system which can mix outdoor and return air to avoid freeze-up problems in northern climates. For example, when preheat coils are used to bring the minimum outside air up to a safe 40°F, 55 percent outside air (now preheated) must be used to provide a 55°F mixture temperature. If only 20 percent minimum outside air were used and preheated to 40°F, a mixture of 67.5°F would result. Mechanical cooling would then be necessary to cool the interior spaces year round. Avoid this situation, if possible.

With a conventional system and the mixed air temperature controlled at 55°F, heating energy must be added whenever the minimum outside air and 75°F return air mixed together produce 55°F air. Let the mixed air controller control the dampers to provide the desired mixed air temperature. This will provide as much as 20 percent outside air for ventilation, even with outside air temperatures of −20°F. With good mixing, this will occur at 25°F outdoor air temperature and 40 percent minimum outside air, 14°F outdoor air temperature and 30 percent minimum outside air, or −20°F and 20 percent outside air.

As a preheat coil discharge temperature is increased, the amount of outdoor air must be increased to reduce the 70° to 80°F return air down to the desired 55°F mixture. Energy waste increases rapidly. More outdoor air must be used and it must be heated to a higher temperature. So use caution when increasing pre-heated outdoor air usage because the energy waste can become tremendous. Such is similarly true for the outside air damper which was, for whatever reason, disconnected to satisfy fresh air complaints during the past "conservation-minded" decade. Better to prudently maintain a good mixing system than to allow knee-jerk reactionary fixes like that.

USING AN ECONOMIZER CYCLE TO REDUCE THE COOLING LOAD

The term "economizer cycle" defines a control sequence which utilizes outside air whenever conditions are such that it provides some "free cooling," thus reducing the cooling load. Digital controls compare temperatures and moisture content of both outside air and building air at all times. This measures the total heat or "enthalpy" of each air stream and chooses the mixture which imposes the least load on the cooling coil.

All devices for sensing moisture in air streams with wide temperature and humidity variations installed prior to 1990 require periodic maintenance. Unmaintained humidity sensors will actually waste enormous energy when they bring in hot, wet air during a failed economizer cycle. Thus, economizer controls should be installed on larger 10,000 CFM air handling systems where proper maintenance can be maintained. However, smaller buildings with packaged rooftop units and sometimes even larger facilities knowing the realities of unpredictable maintenance will use temperature differential economizer cycle. A more reliable, accurate, expensive dry bulb temperature sensor can be used. Or a less expensive approach is to install two temperature sensors (intake and return) electronically programmed into a controller with the average daily humidity for the given geographic climates at various periods of the year.

Modern digital humidity sensors available after 1990 eliminate the concerns for periodic maintenance. This makes it much more practical to conserve energy through standard economizer cycles and enthalpy controls. Also check that the return air dampers tightly seal during economizer cycles. Due to the temperature and pressure differential as well as the fact that this damper was originally designed for mixing only (not total shutoff during 100 percent outdoor air economizer cycles), unexpected leakage can significantly affect energy savings.

CONSERVING ENERGY DURING UNOCCUPIED BUILDING CYCLES

Call it "optimal start," "night setback," "morning warm-up," "optimized start-up and shut down": they all deal with the period of building scheduling known as the unoccupied cycle. The savings in fan horsepower and thermal energy due to unoccupied temperature control, delayed morning start-up, and reduced temperatures at occupancy may be your most significant energy conservation efforts. Unfortunately, none of these can be treated separately. Early shut down affects the energy used during the unoccupied hours at night, and the level of temperatures maintained at night affects the start time in the morning.

Programming of all of the above can range from using timeclocks on air handler fans to a sophisticated building-wide automation computer. Some models of the latter actually use artificial intelligence computers to learn how previous unoccupied cycles worked, and the computer automatically improves the next day's control sequence. Why is all that necessary? Although the basic philosophy of energy savings during the unoccupied cycle is simple, the dynamics of an intelligent building can make the task very time consuming. The most difficult factor to assess is the building's mass temperature. This mass includes the floor, ceiling, and walls as well as high-tech office equipment operating during unoccupied periods. Optimized shut down also requires factoring in the mass temperature of the building occupants and other loads during the occupied period. So even though time clocks can save some energy, this section will describe more of the factors to consider toward maximizing those energy savings while minimizing occupant comfort effects.

CONSERVING ENERGY DURING THE COOLING SEASON
(OPTIMAL PULLDOWN)

The primary cooling equipment is sized to handle latent and sensible loads in an occupied building, the full people and lighting load, the design transmission load, the ventilation load and a full sun load. However, during the pulldown period:

- The people are not present.
- The lights are not illuminated.
- The outdoor air dampers should be closed.
- The sun load is minimal.
- And outdoor temperatures are usually below design.

Although a new load exists (i.e., that associated with cooling down the mass), it is small compared to loads that do not appear at this time. It can be assumed that the primary equipment is sufficiently oversized and the space temperature can be maintained at the level desirable and economical for pulldown. If this is true, you need to examine only the building mass temperature to determine how long before occupancy a system must be started to extract heat from and bring the mass temperature to a desirable level. Outdoor conditions before start-up are not critical factors.

The more tropical climates may use unoccupied humidity override to keep dehumidification under control.

MAKING ADJUSTMENTS DURING THE HEATING SEASON

As the people, lights, and solar radiation add a load to the cooling system in summer, so do they subtract from the load on a heating system.

Heating systems must be sized to handle the building heat losses during the unoccupied cycle. In a building being designed today, it is possible that the system is down-sized to just maintain the building at occupied temperatures during the coldest winter days. In that case, the night setback temperature cannot be any lower than the mass temperature desired at occupancy because the extra warm-up capacity is not available in the system. A system of this size will have to run continuously at the coldest outdoor conditions. Even with this design philosophy, when outdoor temperatures increase, the system becomes more and more oversized, and the philosophy expressed for the cooling season becomes more applicable.

In existing buildings and in new construction where systems are oversized to provide for warm-up requirement, the mass temperature requirement becomes the basic criterion for system start-up. Outdoor conditions become a less important factor if the system is big enough to maintain the desired space temperature all during the warm-up period.

In an existing building or a properly designed new building, it should be obvious that outdoor conditions are not critical in determining the optimal warm-up start time. It should also be obvious that the more the heating system is oversized, the shorter the start-up time can become. This is very important because the major portion of savings is usually in the electrical energy required in the air and water distribution system.

The optimal start program must accurately sense the conditions of the internal building mass temperature to determine how much energy must be added or extracted to bring it to a desired level at occupancy. This is the most important input. The outdoor temperature, solar level, and other factors are not at all important during the pulldown (cooling season) and are of little value during warm-up (heating season)—unless the building has an undersized system. During recent years, some maintenance managers have considered wind velocity and solar intensity as an input by customizing their building automation system computer program. This complication seems to have little or no value on existing buildings and cannot be justified on new construction.

OCCUPIED TEMPERATURE EFFECTS

Proper control of the unoccupied cycle produces savings not only from the decreased heat loss or gain, but more important from energy stored in the building mass during the entire occupied period. The stored energy is given up during the unoccupied period which prevents a rapid change in space temperatures. Not all of this energy is replaced at start-up.

Because one of the areas of savings is the use of mass storage, occupied temperature also affects energy used during the unoccupied cycle. For instance, during the heating season, allow the space temperatures to increase to the maximum economical and comfortable level during the occupied period. This would be especially effective for interior zones normally requiring cooling even in winter. By allowing a slight temperature rise during the last hours of the occupied cycle, (possibly with an early showdown technique described later), extra energy is stored in the building mass. This stored energy can be used to prevent unoccupied temperature reduction, decrease warm-up time, and decrease fan run time. Some control strategists suggest allowing temperatures to drift up to 78°F at higher outdoor temperatures.

If this is done, rarely will any warm-up cycle be required when outdoor temperatures are above 40°F and a good optimal start program is used to enable these dynamic control strategies. Using this philosophy, any building located in a southern state that has less than 300 total hours per year below 40°F, will probably not need to be concerned with unoccupied cycle control during the heating season. Systems should be shutdown, as is normal for the summer unoccupied cycle.

During the cooling season for typical office buildings, systems are usually turned off completely during the unoccupied cycle. Just as no control of night temperatures provide maximum savings in summer, this approach is also practical for the heating season. Although night temperatures are usually not controlled, you should set some low limit (typically around 50°F) to avoid condensation in and on walls. Higher limits may be required if plants or animals are kept in the facility.

Some building automation computers will allow override of shutdown control for evening and weekend use of the building. Via a telephone interface, the occupant listens to a computer voice describe how to easily override temperature, lighting, and other controls by using the telephone keypad. Such override systems are more practical with separately zoned systems like unitary heat pumps rather than large central systems, where a major chiller or boiler must be started to activate one or two zone's air handling system.

A computerized optimal start program will permit the system to be shutdown after occupancy and keep the system off until the last possible moment when it will just be capable of bringing the mass to some predetermined temperature at occupancy time. The predetermined temperature must be based on comfort during the first hour of occupancy weighed against the energy savings desired. This will usually be very close to occupied cooling season temperatures, but at least 5° below the heating season occupied temperature. Only a computerized program can accurately determine weight of a building's mass and then determine the mathematical algorithm for the shape (slope) of the "fall" and "rise" curve of the mass. (See Figure 4-1.)

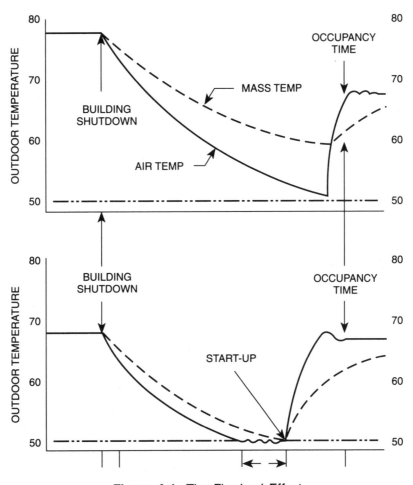

Figure 4-1. The Flywheel Effect

Many optimal start-up programs now include a provision for "early shutdown." If you consider an early shutdown and use the building's "flywheel" effect to permit shutting the system off before the building becomes unoccupied, it is obvious that the temperature during the cold winter season, for example, will drop more rapidly during the unoccupied cycle. This means that the night cycle control may have to turn the fan on sooner and some (or maybe all) of the savings will be lost. Other considerations are whether discomfort at office shutdown will prevent free overtime and whether codes will permit an occupied period without ventilation.

UNDERSTANDING AND MODIFYING AIR HANDLING SYSTEM DESIGN

Historically, mechanical system designers have not been concerned with energy consumption characteristics of their systems. Consequently, constant volume reheat systems which have been the workhorses of the HVAC industry for multistory, multizone buildings, have been utilized at excessive costs in heating and cooling energy. (See Figure 4-2.)

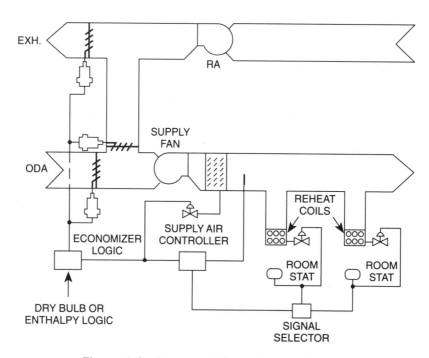

Figure 4-2. Constant Volume Reheat Control

In general, reheat systems have been used with a single zone supply at a nominal air temperature of 55°F. Terminal reheat units provide as many different zones of control as desired. In fact, buildings have been designed with separate reheat coils in all exterior rooms, and with separate coils for several large interior areas per floor.

Reheat systems have very high operating costs because the entire volume of supply air is conditioned to 55°F before delivery to the terminal reheat units, where it is then reheated to satisfy space demands. There is no diversity of cooling load in reheat systems because as internal or external loads decrease, they are artificially replaced by reheat.

One method of conserving energy in reheat systems is to reset the cooling coil discharge temperature upward when cooling loads decrease. By doing this, you reduce cooling energy and decrease the amount of reheat added. Another option involves reconfiguring it into a more modern variable air volume system. (See

Figure 4-3.) A variable-speed drive can be retrofit onto most existing air handler motors. Variable air volume terminal boxes and controls can also be retrofit. This may require a professional controls contractor. One of the most important checks on a modified single duct VAV system with reheat is to assure that the terminal box units reduce to minimum CFM before reheat is energized. But remember that new ASHRAE minimum ventilation standards discourage adjusting VAV boxes to full close, despite possible energy consumption reheating air when at setpoint. Variable air volume systems with reheat should deliver air at the highest possible supply air temperature to satisfy room air conditions and avoid reheat. Ideally, this is done with high signal selectors from representative rooms which reset the supply air when the units are reheating. This will save energy during the period when the system is reheating, and it may save cooling energy at the central system, and fan horsepower. Set supply thermostats lower in free cooling periods and higher in summer for maximum energy conservation.

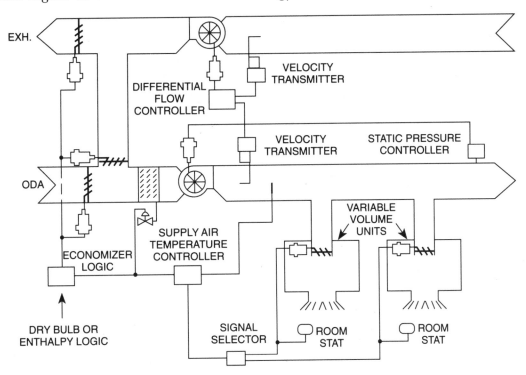

Figure 4-3. Variable Volume Control

You should also control duct pressure to maintain minimum allowable static pressure at terminal units. Regardless of duct sizing, duct pressures vary as flow varies, and pressure at various points in the system vary with respect to each other as loads shift within the building. For this reason, the sensing tip must not be at the fan discharge for maximum economy, but at the lowest point out in the system. Install an easy access hole to regularly clean out dust, which tends to accumulate more at the end of the system, too. If flows vary from one area to another, and this low pressure point shifts, use more than one static tip and controller. Set each controller for the lowest pressure that is required at that point in the duct work. For example, this may mean setting one controller at 1.5 W.G. and another at 2.2 W.G. A signal selector chooses the controller requiring the most output and controls the supply fan accordingly. Locate a high limit static pressure control at the fan discharge to prevent excessive duct pressure in the event a fire damper closes between and the remote sensing location.

SUPPLY AIR RESET TO CONSERVE ENERGY

Figure 4-4 depicts a typical energy-saving supply air reset schedule for a VAV system.

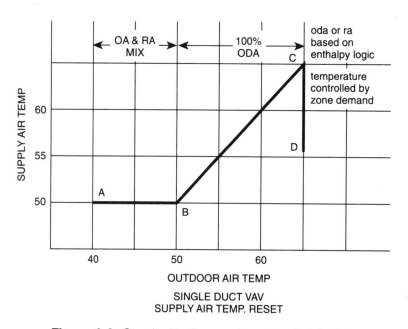

Figure 4-4. Supply Air Temperature Reset Schedule

Point A to point B in Figure 4-4 shows that whenever the outside air temperature is below the minimum acceptable by the diffuser to provide good air distribution, you should use the minimum acceptable temperature. This will maintain fan horsepower at a minimum level. Points B to C indicate that 100 percent outside air is used whenever the outside air is equal to or greater than the minimum acceptable temperature described above. It is used until it is too warm to satisfy the space cooling need before mechanical cooling is turned on. Points C to D show that when mechanical cooling is turned on or made available, two situations exist:

- If the mechanical cooling serves only the VAV system, consider the possible savings by increasing the supply air temperature to increase the chiller efficiency and reduce latent cooling required. It will probably offset the added fan horsepower to deliver more air. However, analyze potential savings before making a decision.

- If chilled water is available at constant temperature. Even under this condition, latent cooling savings may more than offset increased fan horsepower costs. However, analyze savings before making a decision.

The temperature control scheme to reset supply air lower than zone demand when outside air is available for free cooling is rather simple. It merely requires a mixed air controller set for the minimum temperature that can be supplied while still maintaining proper air distribution in the space. An economizer cycle based on outside temperature or enthalpy switchover for maximum economy is used on the outside air dampers to conserve energy at elevated outside temperature conditions. However, this is *not* applicable on VAV systems with reheat operating, because any zone calling for reheat would require more energy to reheat the cooler air.

FILTER REPLACEMENT AND ENERGY CONSERVATION

With increasing indoor air quality concerns, better filter replacement may provide a healthier environment as well as possibly saving energy. Filters perform the vital function of trapping dust particles that would otherwise retard air flow and reduce interior air quality. Inefficient filters also require increased fan power which consumes a surprisingly high percent (5 to 20 percent) of an air conditioning system's total energy consumption.

The effect of dirty filters on the system energy usage will vary with the type of system used. With a heating or cooling system that is cycled on and off to satisfy the load, dirty filters will reduce the air flow through the heat exchangers and conditioned space. The result will be that the equipment is forced to run longer on each cycle. This means increased energy usage due to longer run times and lower efficiencies.

For continuously running constant volume fans, the energy usage of the fan will actually drop as the filter banks load up. This happens because the filters are taking a larger percentage of the total air system pressure drop, and the actual air delivery will decrease. The fan then requires the same amount of energy to move a lower volume of air. Air flow through all the coils will drop below the design values, thus reducing the heat transfer. The conditioned space will not receive the required air volume and proper control will be very difficult to attain.

For variable air volume fan systems, any increase in filter pressure drop will be compensated for through increased fan energy. The system will not suffer from lack of air flow, but the fan horsepower output will increase with filter resistance. A filter during the end of its useful life may require up to five times as much energy as when new.

Indoor air quality studies are showing that fundamental maintenance practices like changing filters sometimes are the leading contributors to this growing phenomenon. Check for filter pressure drops on a bi-weekly basis, and use a manometer instead of simple visual inspection. Filter manufacturers normally catalog a pressure drop rating for clean and dirty filters. The length of time that a filter will perform efficiently depends on atmospheric conditions present. Buildings located in industrial areas may require filter changes every two or three weeks. However, buildings located in relatively "clean" areas may require filter changes on a quarterly basis, which is still more frequent than some managers replace filters.

A computerized building automation system can report dirty filters if pressure sensors are installed after each. Considering the increasing energy-savings for prudent filter replacement as well as indoor air quality concerns, this may be worth the expense if your organization does not diligently follow a preventive maintenance plan.

PHYSICAL PLANT ENERGY SAVING STRATEGIES

You can find good candidates for optimizing the chiller plant in the decreased condenser water supply temperatures and increased chilled water supply temperatures. As a general rule, you can expect the increase in efficiency for centrifugal refrigeration machines to be 1 -1.5 percent for each degree decrease in either water supply temperature. Energy savings will vary based on condenser, evaporator, and overall chiller plant design. Other areas for energy conservation control investigation include chilled water temperature drift, multiple chiller sequencing, dew point control, seasonal scheduling, and cooling tower control.

The greatest potential for savings due to temperature reset of chillers is on the condenser side of the machine. Although the increase in efficiency for a degree of temperature reset is approximately the same for both the evaporator and the condenser sides, the greater amount of reset is usually possible on the condenser side. Chilled water reset typically is limited to 3-5°F with an unlikely maximum of 10°F. However, the condenser water circuit can frequently achieve 10-20°F of reset.

Computerized building automation systems have standard programs for all of the above types of chiller plant optimization. For example, these programs will maximize chiller plant operation by sequencing chillers to keep each loaded in the middle to upper range (40-90%) of their design capacity, where they are most efficient. Most programs will also conveniently produce energy savings reports based on chiller plant optimization.

Energy conservation in boilers starts with good maintenance and upkeep on heating systems. Electronic controls monitoring the flue gas temperature as well as the CO, CO_2 and O_2 contents in the flue gas can increase combustion efficiencies. But because of the nature of boiler operation itself, there are few advantages to sophisticated control sequences like in the chiller plant. Boiler reset of supply water temperature will have little effect on boiler efficiency, and their start-up periods prohibit sequencing.

Peak electrical loads in the physical plant and throughout the building can be a big source of energy savings. Electric utility rates for commercial buildings are typically based on both consumption and demand. The demand charge is determined by peak Kw usage in a specified billing period, sometimes only 15-minute periods during midday in large metropolitan areas. The peak occurs when a surge of electrical loads occurs in one of these periods, such as elevator motors at lunchtime. Once a peak demand charge is established on the electric bill, it may affect rates for many months.

To avoid costly peak surges in electrical usage, modern building automation system computer programs will monitor total Kw usage as well as peak usage. As peak periods like midday approach, the programs will begin shutting off nonessential electrical loads like bathroom fans, closet lights, auxiliary pumps and equipment. It will also schedule start-up loads or prohibit a chiller sequence event until after a peak demand period passes. This basic technique is called *load shedding.*

Another approach used to reduce demand charges is to utilize standby generator sets to peak shave. By operating the emergency generators, electrical demand will be reduced during peak periods. However, with the increasing fuel costs and low cost of electricity in some geographic regions even at peak rates, you should carefully examine the tradeoff costs.

LIGHTING CONSERVATION

Lighting typically represents 30-50 percent of a building's electrical power consumption, or approximately \$0.60/sq. ft./year (which of course varies by geographic region). Because lighting controls can reduce this expense significantly, they are an important factor in a building energy management program. A lighting control system may use several different strategies for reducing energy consumption, these are described in the following sections.

Occupancy-Based Control of Lighting Levels

This is normally the most significant source of energy savings. Typically, the lighting levels are scheduled based on occupancy—i.e., full lighting for normal occupancy, and reduced lighting levels for the cleaning crew. This may be accomplished by simply installing an occupancy sensor in place of wall light switches. Or it could be tied into a computerized building automation system for programmable scheduling. An override function, similar to that for HVAC system override, is typically necessary for after-hours building use.

Daylighting Control

Reducing the "artificial" lighting when natural daylighting is available has a strong emotional appeal. To take full advantage of daylighting, however, the building must be retrofit with several design features. These include building light shelves to enhance daylighting penetration, sometimes up to 20 feet by reflection off the ceiling. The building should also be wired to provide lighting zones 10-20 feet deep running parallel to the windows. For these reasons, daylighting control is normally restricted to new buildings designed for its use.

Tuning

Tuning provides the ability to vary lighting levels to reflect the actual amount needed for the task and age of the occupants. The tuning may occur on a fixture-by-fixture basis or for a large zone. For example, assume that the general overhead lighting has been designed to provide 70fc (footcandles) of lighting throughout the floor. Individual fixture tuning may allow lighting in walkways to be reduced to 20-30fc without affecting the lighting in critical task areas. Large zone tuning, on the other hand, might be used to provide "task-ambient" lighting for a department using VDTs extensively. In this case, the overhead general lighting would be reduced to approximately 30fc, and each occupant would be provided a task light (desk lamp) to provide the 70fc level on the reading area only.

Comparing Various Lighting Control Systems

Daylighting and tuning provide energy savings by reducing the lighting power requirements during normal work hours. They also have a positive impact on building electrical demand charges. Occupancy scheduling, on the other hand, provides savings by reducing the runtime. Because the afterhours reductions do not normally coincide with peak loading periods, electrical demand is not reduced.

Tuning through dimming control requires either split wiring lighting fixtures or retrofit with dimmable solid-state ballasts. In the former, half the fixture or one tube in a three-tube fixture is switched off for moderate level control. Dimmable solid-state ballasts are now available which provide improved efficiency coupled with dimming capability. Another common technique for dimming lights is to replace two fluorescent tubes in a four tube fixture with silver or aluminum reflectors. This provides 75 percent of the light while cutting energy consumption in half! Because most buildings are overlit by design, the lower light levels often improve visual comfort. While installing the reflectors as well as periodically thereafter, clean fixtures and lenses. As lighting is lowered closer to "ideal" levels, regular lens cleaning becomes more important. Relamp fixtures in groups according to a preventive maintenance program based on bulb life, instead of when a bulb burns out. Both the energy savings of not using bulbs during the last quarter of their life as well as the labor savings of getting out the ladder only once far exceeds the small extra cost in bulbs. Used new high-efficiency, energy-saving bulbs whenever possible. For example, fluorescent lamps are more than four times more efficient than incandescents.

Reducing lighting loads will also significantly reduce internal cooling requirements for most buildings. The heat generated by lights, however, is part of the perimeter heating load, which may adversely impact the "breakeven temperature" for the building, if it is reduced significantly. Although the trade-off is well worth it in energy-savings, this still requires consideration during building maintenance and operation.

THE FUTURE OF INTELLIGENT BUILDINGS

Owners and managers of intelligent buildings often experiment with more exotic energy-saving techniques like heat recovery from a building's sewage disposal system, ice storage machines to create chilled water during off-peak electrical demand hours, or even co-generation plants to become totally independent of electric power utilities. Although much of this advanced energy conservation equipment may prove to be both practical and commonplace in the future, those pioneers installing it today probably need little assistance with day-to-day maintenance practices. Nevertheless, keep in mind that you can still realize considerable savings of your utility dollars by improving basic operations and maintenance practices.

Integration is one aspect of intelligence. While energy conservation controls and techniques are being employed, intelligent buildings will integrate the information already being gathered by these systems to:

- Document building operational compliance for regulatory or litigation purposes
- Be a part of the facility management's quality assurance or continuous improvement mechanisms

Such a computer system can also document the energy savings in your building to help show how your conservation measures are paying for themselves.

5

Occupational Safety and Health Regulations

Christopher H. Branton

The Occupational Safety and Health Agency (OSHA) is responsible for providing the employees of all companies with a safe and healthful workplace. There are a few sectors of industry which are not subject to OSHA regulations. However, this is a very difficult matter to prove and it is best to assume that your company falls under OSHA regulations OSHA regulations are can be found in 29 Code of Federal Regulations (CFR) Part 1910. Additional regulations which govern the construction industry are located in 29 CFR 1926. This chapter attempts to provide the highlights of portions of the regulations which apply to the building maintenance industry. Remember that these regulations carry the weight of the law and that violators are subject to fines of up to $70,000 per violation. In addition, OSHA and other regulatory agencies have been increasingly asking for, and receiving criminal penalties for employers that willingly violate OSHA standards.

OSHA regulates many other areas which may be of concern to your facility; however, this chapter covers only the regulations that are most relevant to maintenance management. Additional topics include fall protection, lighting requirements, and specific chemical regulations, to name a few.

If you have questions regarding other safety and health procedures, contact the corporate safety and health representative or industrial hygienist within your company. If one is not available, contact professional safety consultants and industrial hygienists in your area.

COMPLYING WITH PERMIT REQUIRED CONFINED SPACE STANDARDS

OSHA regulates confined spaces under 29 CFR 1910.146, the Permit Required Confined Space Standard. The standard became effective on April 15, 1993. The definition of confined spaces under the standard is slightly different than that used in the past. A space is considered confined if it meets the following criteria.

- It is large enough an so configured that an employee can bodily enter the space

- It is not designed for continuous occupancy.

Once it has been determined that a space is confined, it must be determined whether a permit is required for the space. If any of the following criteria are met, the space is considered a permit required confined space, and is subject to all provisions of the standard:

- The space contains, of has the potential to contain a hazardous atmosphere

- The space contains a material that poses an engulfment threat to entrants

- The space has an internal configuration that could trap or asphyxiate an entrant by inwardly converging walls or by a floor which slopes downward and tapers to a smaller cross-section

A hazardous atmosphere is an atmosphere that may expose employees to the risk of injury to their health, and includes any of the following:

- A flammable gas, vapor, or mist in excess of 10% of its Lower Flammable Limit (LFL).

- Airborne combustible dust at a concentration that exceeds its LFL

- Atmospheric oxygen below 19.5% or above 23.5%

- Atmospheric concentration of any substance for which a dose or permissible exposure limit (PEL) has been established that could result in employee exposure above the limit.

- Any other atmospheric condition that is immediately dangerous to life or health

Most people are aware of the classic confined spaces, such as tanks, manholes, and sewers. However, most people are unaware of the nonclassic spaces. These include pits, some pump rooms, excavations over four feet, and retention berms surrounding tank farms.

Ensuring Safe Entry into Permit Required Confined Spaces

Before entering a confined space, maintenance workers must follow a number of procedures. These procedures must be included in a confined space entry permit (See Figure 5-1). This permit is not only required for entry, but also provide

---

CONFINED SPACE ENTRY PERMIT

Location: **Date:**
                                        Time:

Purpose for Entry:

Personnel on Site: **Designation:**

Emergency Contacts (telephone numbers):

Plant Personnel   _____   Fire: _____
Hospital: _____   Police: _____

Equipment (check where appropriate):

___	Lockout/Tagout	___	Respirator (type: _____)
___	Lines Broken/Capped	___	Lighting (type: _____)
___	Purged	___	Harness
___	Ventilation	___	Lifeline
___	Area Secure	___	Tripod
___	Protective Clothing (type: _____)	___	Fire Extinguisher

Atmospheric Testing Performed:

Test	Entry Limit	Test 1	Test 2	Test 3	Test 4
% Oxygen	19.5 to 23.0%	___	___	___	___
% LEL	> 10%	___	___	___	___
Hydrogen Sulfide	> 5 ppm	___	___	___	___
Carbon Monoxide	> 25 ppm	___	___	___	___
Others:					
_____		___	___	___	___
_____		___	___	___	___

Testing Equipment:

Instrument:	*ID #:*	*Calibration Date:*
_____	_____	_____
_____	_____	_____

Authorized Signatures:

Supervisor:	_____	_____
Others:	_____	_____
	_____	_____

Figure 5-1

the entrant with a checklist to ensure that all components of the program are being followed.

Make sure the area is completely isolated before entering the space. This includes:

- Locking out or disconnecting all electrical and mechanical connections
- Locking out and blanking all piping
- Clearly marking the area and alerting all personnel in the area

Before entering the space, you must also ensure that a hazardous atmosphere, as defined above, is not present. A number of instruments are available to monitor confined space atmosphere. An instrument should monitor for at least the following:

- Lower Explosive Limits
- Oxygen Content
- Hydrogen Sulfide

The recommended limit for entry for entry without protective equipment (respirators/protective suits), are under 10% LEL, 19.5% oxygen, and 5 parts per million (ppm) of hydrogen sulfide. In addition, you must also monitor for any other potential contaminants that may be in the space. Instruments are also available to provide direct readings of concentrations on almost any contaminant you many encounter. Take readings in at least three location within the space (top, middle, bottom). This will ensure that contaminants are not collecting in a specific area of the space. If you encounter a hazardous atmosphere do not enter the space without proper protective equipment. The area should be ventilated to reduce the amount of contamination to acceptable levels.

All personnel that participate in confined space entry procedures must be trained properly. In general training will include the following:

- Proper use of monitoring, ventilation, communication, personal protective, and rescue equipment.
- Proper operation of lighting, barriers, and ladders.
- Communications without outside personnel.
- Evacuation procedures.

This is the minimum training required to enter a permit required confined space. Additional training is required for attendants that provide the entrants link to the outside. Attendants may not leave the entrance to the space at any time. Attendants must also keep track of all entrants and be familiar with procedures to contact rescue personnel. The attendant is not normally part of the rescue team and may not enter the space for rescue purposes. The attendant, must also be certified in basic first aid and cardiopulmonary resuscitation (CPR).

Entry supervisors must also be trained to authorize permits, and determine when the entry has been completed. Supervisors must have the authority to stop entries at any time if they determine that the space is unsafe.

Rescue personnel must receive all the training listed above with the addition of the annual hands-on drill that simulates an entry rescue. Rescue teams must also have specialized equipment such as SCBAs, and extraction equipment.

Confined Space Rescue

Confined space rescue personnel must be immediately available during any entries. An in-house team may be used if they are properly trained and supplied as described above.

If an outside rescue service is to be used, they should be contacted prior to any entries to verify availability and response time. If an outside rescue team is used, all efforts must be made to facilitate non-entry rescues by on-site personnel. This can be accomplished by having all entrants wear a harness with an attached lifeline. In spaces that would require personnel to be pulled out more than 5 vertical feet, a mechanical device should be used.

COMPLYING WITH REGULATIONS ON HAZARD COMMUNICATION (RIGHT TO KNOW LAWS)

OSHA has issued a standard (29 CFR 1910.1200) known as the Hazard Communication Standard or Right-To-Know Law. This law is designed to keep employees informed as to the potential hazards that exist in the workplace. The employer must develop a written hazard communication program and provide adequate training for employees. This section addresses the major points which must be addressed in your hazard communication program.

The Hazard Communication Standard should not be confused with the Community Right-To-Know Regulations that were developed by the U.S. Environmental Protection Agency (USEPA). These regulations are generally carried out on a state or county basis and allow for emergency and disaster planning in your area. Contact your Local Emergency Planning Commission (LEPC) or State Emergency Planning Commission for more information on this program.

Labeling Containers

All containers containing hazardous materials must be properly labeled. The most popular labeling system has been developed by the National Fire Protection Agency (NFPA) and is commonly known as the safety diamond (see Figure 5-2). The diamond provides information on flammability, toxicity and, reactivity of chemicals. Additional space is provided for special hazardous properties. All ingredients are rated on a scale of 0 to 4, with 0 with the least dangerous and 4 the most dangerous. Most products arrive with these labels in place; however, you should have a supply of labels available for any containers that arrive improperly labeled. REMEMBER: Even small containers which may be filled from larger ones (i.e., a spray bottle filled from a drum of cleaning agent) must be properly labeled.

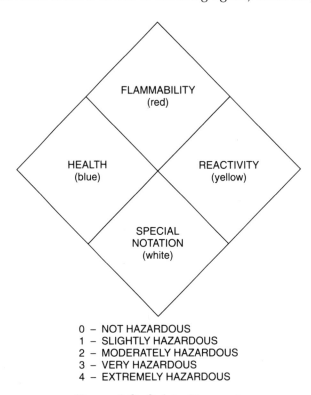

```
0 – NOT HAZARDOUS
1 – SLIGHTLY HAZARDOUS
2 – MODERATELY HAZARDOUS
3 – VERY HAZARDOUS
4 – EXTREMELY HAZARDOUS
```

Figure 5-2. Safety Diamond

Using Material Safety Data Sheets

Make sure Materials Safety Data Sheets (MSDSs) are available to employees for inspection. The MSDS provides important information that employees should refer to in the event of a spill or employee exposure to a chemical agent. An MSDS should provide information on the following:

- Chemical identity and ingredients
- Physical and reactivity properties
- Fire fighting procedures
- Storage requirements
- Health effects
- First aid procedures
- Spill procedures
- Protective equipment
- Special precautions for handling

Most products arrive with an MSDS provided in the packaging. If one is not included, you should write to or call the manufacturer. The manufacturer is required by law to provide you with an MSDS for all products.

Keep a complete list of all materials and their location along with appropriate MSDSs in an area where all employees have access to them. If there are a number of areas which contain hazardous chemicals, then keep a smaller list these areas.

Training Employees in Hazard Awareness

All employees must receive training regarding the companies Hazard Communication Program. This training must include the locations of hazardous materials and MSDSs. Conduct training in the use of MSDSs and the information they contain. Have employees provide you with the MSDS and location of any new materials which enter your facility. This will allow you to keep your hazard communication program up to date.

UNDERSTANDING CHEMICAL AND PHYSICAL PROPERTIES OF MATERIALS TO ENSURE SAFE HANDLING

Thousands of chemicals are used in industry. Each chemical is defined by its chemical and physical properties, including boiling point, corrosivity, toxicity, and density, among others. These properties may place a chemical into one of the following hazardous materials classifications:

- The material represents a *fire hazard*. It may be explosive, flammable, or undergo a reaction that gives off heat (exothermic),

- The material is *cryogenic*. It presents a hazard due to its extremely cold physical state.

- The material is *toxic*. Depending on the dose, this material will cause sickness or death.

- The material is *radioactive*. These materials react with human tissues to cause cell death or mutation.

A basic knowledge of the chemical and physical properties of materials helps your staff handle and store the materials in a safer manner. It also helps workers deal with the materials more confidently if they should be spilled. This section is intended to familiarize you with the terms and concepts used to define the physical and chemical properties of materials which may be present in your building. Information specific to materials in your building should be available in the MSDSs for a particular substance.

Recognizing the Physical State of Materials

There are three physical states that a material can take. This physical state can be directly related to the danger posed by a hazardous material.

- Solids—maintain their own shape, such as wood and metal. Common types of solids include:
  - Dusts—generated by grinding or crushing solid materials.
  - Fumes—solid particles which have condensed from a gaseous state. Welding fumes are the most common.
  - Smoke—the byproduct of incomplete combustion. Tobacco smoke contains tarry particles and is classified as a wet smoke.
- Liquid—a material that takes the shape of the container it occupies. Common liquids include mists: small liquid particles generated through spraying or dispersion. Spray painting creates a mist.
- Gas—a formless fluid which expands to completely fill the space it occupies. Industrial gases are often held under pressure in cylinders. Vapor is the gaseous form of a substance which is normally a liquid or solid at room temperature.

Recognizing Physical Properties of Industrial Materials

The physical properties of a material help describe a materials appearance. All Material Safety Data Sheets (MSDSs) contain information about a materials physical properties:

- Density—mass per unit volume of a solid or liquid [i.e., grams (g) per milliliter (ml)]. The density of water is 1.0 g/ml.
- Vapor density—mass per unit volume of a gas. The vapor density of air is 1.
- Specific gravity—relative measure of density. It is the ratio of the density of a material to water.

Vapor density and specific gravity are very important when dealing with environmental pollutants. These physical properties determine how a material travels through the environment and precautions which may be necessary for personal protection and methods of containment.

Solubility is another important physical property of a material. This property is generally expressed as the weight or volume of a material which will dissolve in water. Water is used because it is the most common solvent (the material that another material is dissolved in). Solubility is an important property to know if a spilled material is on or near a waterway. Solubility of a material is also important when selecting decontamination solutions.

The potential for fire and explosion are common hazards that are associated with many hazardous substances. Although it may not always be obvious, a flammable liquid does not actually burn. Instead, the liquid gives off flammable vapors that ignite when a flammable concentration has been reached. The following terms are important to understand to help deal with flammable or explosive hazards:

- Vapor pressure—the pressure exerted when a liquid is in equilibrium with its own vapor. Vapor pressure depends on both temperature and pressure.
- Combustion—a chemical reaction which takes place between two substances. One of these substances is normally oxygen. Combustion normally produces heat and light. Complete combustion is a rare occurrence and byproducts often include smoke, carbon monoxide (CO), and other toxic gases.

For fire to be sustained three elements must be present: heat, fuel, and oxygen. Removing any of these elements extinguishes the fire. This is the basis for fire extinguishers. Most fire extinguishers eliminate oxygen from a fire to extinguish it. Extinguishers can be placed into four groups depending on the type of fire that they may be used to extinguish:

- Class A—ordinary combustibles (wood and paper)
- Class B—flammable liquids (grease, gasoline, and oil)
- Class C—electrical equipment
- Class D—combustible metals

The most common extinguisher is the ABC extinguisher which uses a dry chemical to coat the burning material and deprive it of oxygen.

Even when all of the components for a fire are present, an ignition source is still necessary. Ignition sources may be sparks, open flames, or heated sources. This is the reason for choosing nonsparking tools when working with potentially combustible materials.

A number of terms are used to describe a materials combustible properties:

- Lower explosive limit (LEL)—the minimum concentration of gas or vapor in air below which combustion will not occur
- Upper explosive limit (UEL)—the maximum concentration of gas or vapor above which combustion will not occur
- Flammable range—the difference between the UEL and the LEL
- Flash point—the minimum liquid temperature at which an ignition source will cause a flash over the vapor level above a liquid.

Recognizing the Chemical Properties of Industrial Materials

The chemical properties of a material describe the chemical class and define such properties as reactivity and corrosivity. The chemical properties define the compatibility or incompatibility of materials that can help you decide how and where they should be stored. There are two general classes of compounds:

- Organic compounds which contain carbon
- Inorganic compounds which do not contain carbon

One of these subdivisions for organic compounds is that of polymers. These are high molecular weight organic materials made up of repeating units. Polymers can be processed into films, sheets, fibers and coatings and are found in many aspects of everyday life. They may be chemically inert but will melt or burn upon heating. The smoke that evolves from burning organic or polymeric materials can be toxic.

Organic and inorganic compounds may be further divided into several classes. For example, these include:

- Corrosives—defined by the Resource Conservation and Recovery Act (RCRA) as having one of the following properties:
  - A pH less than or equal to 2, or greater than or equal to 12.5; or
  - The ability to cause the corrosion of steel at a rate greater than 1/4 inch per year (as determined by industry testing procedures).

  Corrosives can destroy metals as well as organic materials (such as skin tissue).

- Oxidizers—materials that can cause ignition, combustion or detonation of organic materials, powdered metals and other reducing agents. Oxidizers are generally used as bleaching agents found in cleaners, fertilizers and the like.

- Cryogens—gases that must be cooled to less than −150° F to achieve liquifaction. The handling and storage of such materials requires special attention.

UNDERSTANDING THE TOXIC PROPERTIES OF MATERIALS TO ENSURE SAFE HANDLING

To identify health risks that may be present in your facility, you should understand the basics of toxicology. Although this section does not go into the health affects of specific chemicals, it should help you better understand the potential health affects with the aid of an MSDS. All substances have the ability to produce toxic effects. It is the amount of the substance that causes a toxic response. Water may be toxic if large enough quantities are consumed, or it is breathed into the lungs.

Toxicology is the study of the effects of chemicals on living organisms. The toxicologist is trained to examine the nature of the adverse effects of the chemicals and determine the probability of their occurrence. Thousands of new chemicals undergo toxicological evaluation each year to determine their effects on man.

Toxicology has its own vocabulary which we will attempt to explain. In addition, you will become acquainted with the general principles of toxicology.

Understanding Exposure to Toxic Materials

A hazardous substance may enter the body and cause a toxic effect through three basic routes. They are: inhalation, absorption, and ingestion. The most common route of exposure is absorption of chemicals, which generally results in contact dermatitis (skin rash). The most serious route of entry is generally inhalation. Ingestion can also be very serious, however, it is not a common cause of exposure in general industry.

Measuring Toxic Effects

The dose-response relationship is the basis for measuring all toxic effects. There are three assumptions which are made in dose-response relationships:

- The response observed is due to the chemical administered.
- The response is related to the dose or the amount administered.
  - There is a receptor site that acts to produce the response.
  - The response depends on the concentration at the receptor site.
  - The concentration is related to the dose.

- There is a quantifiable method for measuring the toxicity of a chemical.

Therefore, increasing the dose should increase the toxic response.

Toxic responses are varied in nature and can range from a mild rash to death. Some basic types of responses are:

- Acute response—respond in less than 24 hours
- Chronic response—respond in more than 24 hours
- Reversible response—the body is able to repair damage caused by toxin
- Irreversible response—the body is permanently damaged by the toxin
- Local response—occurs at the site of contact (i.e. dermatitis).
- Systemic response—requires that the toxin enters the body and reaches a target organ to cause a response

There are many instances when an individual may be exposed to a number of chemicals at the same time. When this happens the chemicals may interact in one of three different ways:

- Additive effect—the sum of the individual effects of the chemicals.
- Synergistic Effect—of two chemicals combined is greater than the two substances would be individually.

• Antagonistic Effect—two chemicals act against one another and reduce the effect that the chemicals would have individually.

USING PERSONAL PROTECTIVE EQUIPMENT TO ENSURE WORKER SAFETY

To protect yourself and your staff from potentially harmful effects of chemicals, it may be necessary to wear personal protective equipment. This equipment may range from safety shoes and hard hats to a totally encapsulating suit and self-contained breathing apparatus (SCBA) for spill response. The most important concept to understand is that no one type of protective equipment is effective against all chemicals. It is necessary to select different types of protection for each situation that you encounter.

Protective Clothing

The most common occupational illness is dermatitis due to skin contact with irritating chemicals. The most likely part of the body to come into contact is the hands; therefore, one of the most common pieces of protective equipment is the glove. Other common types of protective clothing are boots and coveralls. All of these types of clothing are available in numerous materials and configurations. To select the proper clothing for your job, consult the manufacturer or your safety and health officer. Manufacturers often provide technical assistance in selecting the proper equipment.

Equipment for Respiratory Protection

If there is a potential for exposure to airborne contaminants, you may need to wear respiratory equipment. There are two general types of respirators. The first is the air-purifying respirator, which generally consists of a cartridge through which air is drawn. The cartridge filters out contaminants before they enter the body. Cartridges are available for many types of compounds; however, there is no cartridge that is effective against all contaminants, and it is necessary to evaluate the type of cartridge you need for each situation. These cartridges cannot supply oxygen and may not be used in areas where there may be an oxygen deficiency.

The second type of respirator is the air-supplying respirator, which involves breathing bottled air. The bottles may be located in a remote location or they may be carried on the back as in SCBAs. These types of respirators limit the maneuverability and the time that a person can operate before running out of air. You should always be aware of the amount of air remaining in your system or have a standby person available to alert you when the level of air is getting low. If you are purchasing bottled air, make sure that it is certified grade D, breathable air. Do not use an air compressor for breathing air unless it has been designed specifically for that purpose.

PREVENTING HEARING IMPAIRMENT

One of the most common occupational illnesses is that of hearing impairment. Hearing impairment results in decreased productivity, as well as quality of life. Noise exposure is regulated by OSHA under 29 CFR 1910.95. This standard requires that you monitor all high noise areas to determine the exact amount of noise that is in an area.

Monitoring High Noise Areas

OSHA has determined that noise exposure may not exceed 85 A-weighted decibels (dBA) over an eight-hour work day. As a good rule-of-thumb, any area where you must raise your voice to be heard exceeds the 85 dBA limit. Common areas where noise exposures may occur are:

- Boiler rooms
- Generator rooms
- Near power-saws and drills

Through wear and tear on belts and chains, many pieces of equipment that did not present noise problems in the past may become problems. Therefore, a good preventive maintenance program helps prevent hearing loss to employees.

Eliminating Noise

You can prevent employees from being overexposed to noise in several ways. The most common is the use of hearing protection, which can take a variety of forms—from ear muffs to specially designed ear plugs. Check with the manufacturer that the protection you require is met by the type of hearing protection you purchase. Currently, one of the best hearing protection devices on the market is a foam plug which expands when placed in the ear canal. This device provides a good seal to prevent noise from leaking in around the sides of the devices.

Ideally, you want to eliminate the source of the unwanted noise completely. This may be as simple as changing a belt which is worn and squeaky, or it may involve placing acoustical absorbing materials around the area of concern. This type of work is best done by a professional such as an acoustical engineer.

PREVENTING AND CLEANING UP RELEASES OF HAZARDOUS MATERIALS

There is always the possibility that a hazardous material stored or used within the building may be released. These materials include everyday cleaning materials, chemicals that are used by companies leasing space in your building, or everyday materials that are used within the building such as heating oil. By definition, a release includes any spill, leak, discharge, or other act which causes the uncontrolled introduction of materials to the environment. In the event that you have a release of a material exceeding the reportable quantity of that material, you are required by law to contact the National Response Center at the following number:

National Response Center: 1-800-424-8802

The center is staffed 24 hours per day and offers advice and assistance in clean-up operations. Another useful resource is Chemtrec, which is funded by donations from chemical manufacturers and offers technical advice and assistance in the clean-up of spills. Chemtrec can be reached 24 hours per day at the following phone number:

Chemtrec: 1-800-424-9300

Recognizing Hazards

The best way to both eliminate and recognize hazardous releases is to inventory all potential areas in which a release may occur. The first place to start is outside the building. Outdoor releases may occur at the following locations:

- Outdoor and underground storage tanks
- Bulk materials fill-pipes
- Outdoor storage areas (sheds and storage pads)
- Air conditioning and refrigeration units
- Transformers

After you have identified all potential release sites outside the building, inventory the inside. Inside areas of concern include:

- All piping which contains hazardous material (including fuel lines and heat exchangers)
- Storage closets and rooms (check for type and size of containers)
- Small laboratories or process areas (including dyeing and dry cleaning areas)
- Xerography and blueprinting machines and their associated chemicals

Be aware of these areas and the types of releases that could occur in each of them.

Training Employees to Handle On-Site Spills

To respond to any hazardous materials spill you must have proper training. This training is required by the OSHA regulations found in CFR 29 1910.120. We will limit this discussion to training required for employees who respond to on-site spills. These employees require 24-hours of safety and health training before they can participate in emergency response operations. As a minimum, this training must contain the following components:

- Chemical and Physical Properties
- Hazard Recognition and Analysis
- Toxicology
- Air Monitoring
- Medical Surveillance
- Personal Protective Equipment
- Site Controls
- Decontamination Procedures

A well-constructed course will also include practice sessions which allow employees to practice their skills, as well as become aware of any insufficiencies their response plan may have. These courses are commercially available from many companies, and you should review several before selecting the one that best suits your facility. It is best to train a small number of personnel which will make up your spill team. In general, at least five persons should be available to respond to a spill at any time.

Many small facilities are unable to afford the costs involved with training a spill response team. If this is the case at your facility, investigate hazardous waste management companies in your area so that you can contact them in the event of a spill. These companies will provide appropriately trained personnel and the proper equipment to deal with almost any situation you may encounter. Present the company with a situation and a time frame that may occur at your facility and ask them to tell you how they would respond. If the response makes sense and you are satisfied with the answers to your questions, keep their phone number available in case of emergency.

Containing Hazardous Materials Releases

If you should come across a release of hazardous materials in any of the areas of your building, the first object is to prevent further damage to the building, and injury to personnel in the area. This may be as simple as sounding the fire alarm and evacuating all personnel from the building. When in doubt as to the extent or nature of the spill, this is the best action.

You may be able to stop a spill by turning a valve at a remote location, or uprighting a container. It is important to know the location of all valves and switches in your building so that you can stop a spill and prevent further damage to the environment and personnel. If you are unable to control a spill from a remote location and have not received proper training, you should evacuate the area and contact properly trained personnel.

DECONTAMINATING YOUR FACILITY

Decontamination is the process of removing contaminants from surfaces. The actual decontamination may involve physical or chemical removal. The most common decontamination solution is water, which may be used to wash away any contamination that has contacted the body. If you are dealing with compounds that are not soluble in water (i.e., heavy oils), you may need to use an organic solvent such as diesel fuel. Never allow decontamination solutions to enter drains or sewers unless you are sure that the material is nonhazardous and your local sewer authority can treat it properly. Again, decontamination areas where a spill has occurred should only be carried out by properly trained personnel.

DISPOSING OF HAZARDOUS MATERIALS

As stated, never allow materials to enter drains or sewers until you have made sure that your sewer system is able to handle the material. Any material that is classified as a hazardous waste must be dealt with in a special manner. Federal, state, and local regulations will govern how you must handle the waste material. Hazardous waste may be liquids or solids that were spilled, or protective clothing that became contaminated during clean-up operations. To ensure that you are protected you best bet is to deal with a reputable hazardous waste management firm.

To generate or dispose of a hazardous waste you must have a USEPA Hazardous Waste Generator I.D. Number. This number system allows the USEPA and state to monitor the generation and disposal of hazardous wastes. Your state environmental agency will be able to provide you with the necessary applications to obtain a generator number. In addition, all vehicles that carry hazardous wastes must be licensed hazardous waste haulers in all states in which they will travel. Disposal facilities, including treatment plants, incinerators, and landfills, must also be certified by the federal, state, and local authorities. Ask to see all appropriate certifications before you allow wastes to leave your facility.

To prevent the "midnight-haulers" that dumped hazardous wastes along the side of the road in the past, a manifesting system has been developed to track hazardous wastes at all times. The manifest consists of a form which is unique to each state and must accompany the waste from the time it leaves the generating facility until it reaches the disposal site. Each person or company that comes into contact with the waste must complete a section of the manifest. Be sure to get a copy of the completed manifest for your records to ensure that the waste was disposed of properly. In addition, if your waste has gone to an incinerator, be sure to obtain a certificate of destruction, verifying that the waste was incinerated. A complete list of all hazardous wastes that have left your facility must be filed with the USEPA every two years.

Keeping the proper records will allow you to know the exact quantity and location of all waste that has been generated on your site. This may be important if one of the disposal facilities you have sent waste to becomes an environmental problem. In the event that this happens, the USEPA will attempt to determine the potentially responsible parties and seek money from them to carry out clean-up operations. There have been cases where one drum label has been sufficient to name a company as a responsible party. If you use hazardous waste management firms with a good reputation, you may pay a higher price for disposal, but you will be assured of protection of paying an even higher price in the future.

MANAGING ASBESTOS PROBLEMS IN BUILDING

Asbestos is a naturally occurring silica fiber. Asbestos has a large capacity to absorb heat. Therefore, it serves as an excellent insulator. It can be woven into fabrics which are used as insulating barriers. It has also been used in the production of numerous building materials, including insulations, ceiling tiles, cement, floor tiles, and adhesives. Asbestos is also used in valve seal packing, refractory bricks, and break materials. This section covers problems with asbestos-containing building materials (ACBM).

Asbestos fibers are small enough to enter the lung and reach the alveolar level, where gas exchange takes place. Once in this area, the fibers maintain their shape and are not broken down by the bodies natural defenses. The result is scarring or fibrosis of the lung. These scars can be seen on x-rays as plaques and decrease the ability of the lung to provide oxygen to the body. Severe scarring results in a breathing disorder similar to emphysema, known as asbestosis.

In addition to asbestosis, a unique kind of lung cancer, mesothelioma, is also found in persons exposed to asbestos. Mesothelioma is a cancer of the outer lining of the lungs. It normally takes 20 or 30 years for cancer to develop in individuals exposed to asbestos.

If your facility is a primary or secondary school, you should already be familiar with the inspection process mandated by the USEPA. However, general industry and office buildings may also require an asbestos inspection. First, walk through your own building and attempt to identify any areas that you believe may contain asbestos. Pay particular attention to steam pipes and boilers. Do not attempt to sample the potential ACBM yourself, unless you have received the proper training.

If you believe that you may have asbestos in your facility, have an inspection performed. A number of firms devoted entirely to asbestos sampling have come into existence since the regulations governing asbestos came into effect. Although it is not always required, make sure that the person conducting your survey is a certified USEPA inspector. The inspector will remove small pieces of the suspect material for laboratory analysis. The laboratory will be able to identify the material as non-asbestos or asbestos containing. The laboratory will also be able to identify the type of asbestos involved.

If you determine that asbestos is present in your facility, you may have the ACBM removed, or you may leave it in place. Material that is badly damaged or friable poses a health threat and should be removed. Sprayed-on insulation and pipes wrapped with woven insulation are more likely to become friable than other ACBMs. If you choose to have the ACBM removed, use a qualified asbestos abatement firm. The abatement process involves completely enclosing the contaminated area where removal will take place and keeping it under negative pressure to keep fibers from escaping from the area. In addition, and if possible, the ACBM should be wetted with a water solution to keep the amount of free fibers generated during the removal activity to a minimum.

Once the removal process is complete, keep the area enclosed until air sampling results indicate that no asbestos fibers remain airborne. Send the waste generated during the abatement procedures to a sanitary landfill for disposal.

Asbestos removal can be costly and may pose an increased health risk due to the fibers which may be generated during removal procedures. Therefore, you may choose to encapsulate the ACBM in your area. This is generally a good idea if the ACBM is in an out-of-the-way area when it is unlikely to be disturbed. Encapsulation involves placing a protective coating over the ACBM. A number of commercial encapsulants are available depending on the application. Again, this work should be carried out only by personnel trained in the handling of asbestos. Periodic inspections of encapsulated asbestos material should be conducted and documented. In addition, make sure all employees are aware of its location to avoid disturbing the material and creating a health risk.

MANAGING HEALTH PROBLEMS CAUSED BY POLYCHLORINATED BIPHENYLS (PCBs)

Polychlorinated biphenyls (PCB) are extremely stable, nonflammable compounds that have been used chiefly in the electrical industry as an insulator in capacitors and transformers. They are also used in the manufacture of plasticizers, hydraulic fluids, inks, sealants, adhesives, pesticide extenders, and microencapsulation of dies in carbonless duplicating paper. The U.S. Congress outlawed the production and sale of PCBs in 1976. PCBs are still being used but are slowly being phased out of most operations. All personnel that deal with PCBs should receive training specific to the materials they will be handling,.

Skin contact with PCBs can result in irritation ranging from a mild rash to large cysts. PCBs are absorbed into the body through the intact skin. Once inside the body they primarily attack the liver. Symptoms of PCB exposure include jaundice (yellowing of the skin), nausea, and vomiting. There is limited evidence that PCBs may cause cancer of the liver.

When PCBs are burned or heated to decomposition, as in an electrical transformer fire, they form dioxins. Dioxins were common herbicides contaminants until it was determined that they were human carcinogens.

Complying with Regulations for PCBs

The major regulations which you will be concerned with are those governing PCB transformers. These regulations are part of the USEPAs Toxic Substance Control Act (TSCA). The USEPA defined a PCB transformer as any transformer containing oils with a concentration of PCBs in excess of 500 parts per million. Four types of transformers have been identified: network-high, network-low, radial-high, and radial-low. A network transformer is part of a group of transformers that act in a group. A radial transformer is the only transformer in a specific line. Low voltage is any voltage below 480 Volts (V).

All PCB transformers within commercial areas should have been phased out by October 1, 1990. A commercial area any area with public or employee access, and a 30-meter zone surrounding such an area. The one exception to this regulation is the network-low transformer which may continue to be used until 1993 before being phased out. If you still have or suspect that you have PCB transformers at your facility, have them tested.

Testing for PCBs

To determine if transformers or other containers at your facility contain PCBs, you have two options. The first is to perform on-site analysis. A number of relatively inexpensive kits are available for this purpose. They generally involve removing a small quantity of the material in question, and adding a color indicator to the material. If you plan to do this yourself, you should receive proper training in protective equipment use. These tests serve as a way of determining if PCBs are present and are generally accurate to within 25 percent of the actual concentrations present.

If you determine that PCBs are present in the material tested, a laboratory analysis should be performed. The laboratory will be able to provide a more accurate result, which you will be able to use to determine if your transformer or other system is below the 500 ppm criteria. Nearly all environmental laboratories are able to perform this analysis.

Abatement Options

Prior to the October 1, 1990 deadline, it was possible to drain and flush PCB transformers until they contained less than 500 ppm of PCBS. The transformers could then be refilled with mineral oils and put back into use. You may still attempt to petition the USEPA to retrofill your transformer; however, it is unlikely that they will agree to this. This method is still viable for other types of equipment such as heat exchangers and hydraulic lines.

It is also possible to protect radial transformers with surge protector. This also involves petitioning the USEPA and it is likely that radial transformers with surge protectors will also be phased out at some later date.

The final option, decommissioning, applies not only to transformers, but also many other PCB containing systems. Decommissioning involves draining all PCB containing liquids into drums or tank trucks, and then dismantling and removing the contaminated equipment. Contaminated equipment should only be transported on specialized flat-bed trailers fitted with a three-inch lip to prevent leakage.

Disposing of PCBs

Equipment contaminated with PCBs should be treated as a hazardous waste and you should follow all of the requirements of hazardous waste shipping. Generally, this type of material will be accepted by a hazardous waste landfill. The hazardous waste management company that performs the decommissioning should be able to locate the appropriate landfill and provide hauling to the site.

PCB-contaminated liquids should be incinerated to avoid any potential liability in the future. PCB incinerators are dedicated to PCBs only and do not handle other types of waste. They must meet a number of requirements set forth by the USEPA, including 99.9999 percent destruction of materials, and a certain operating temperature (1200°C or 1600°C depending on the time in the incinerator). Make sure you receive a copy of the manifest stating that the materials arrived at the incinerator, as well as a certificate of destruction from the incinerator verifying that your wastes were destroyed.

MONITORING RADON

Radon is a naturally occurring radioactive gas which makes up a portion of what is commonly called background radiation. It has been found in every state in the country; however, elevated concentrations have been found in the mid-Atlantic, Southern, and mid-Western states. Radon is emitted from soils and has been found to accumulate in basements with poor ventilation. Although this is the most common area for radon to accumulate, it may also be found in offices and other buildings.

Radioactive compounds are capable of producing cancer and genetic changes in humans. Radon acts primarily on the lungs, due to the fact that it is a gas which can be inhaled. A majority of the lung cancer cases not attributable to tobacco smoke in the U.S. may be due to radon. However, research has also been conducted that demonstrates that radon has little or no effect on the production of cancer. Additional research is being conducted by the Department of Energy to determine the exact effect of radon on human health.

You can not see or smell radon gas, and the effects of radon on the body are not apparent until it is too late. Therefore, if you have a basement located in a high radon area, you should monitor for radon. A number of kits are commercially available at discount and hardware stores. The kits are relatively inexpensive and are easily operated.

If you find that you have elevated levels of radon, install a ventilation system which will not allow radon to accumulate in the area of concern. This system should be designed by an HVAC expert to ensure that the system is adequate to eliminate the risk to employees.

MONITORING INDOOR AIR QUALITY

There has been an increase of concern regarding indoor air quality over the past few years. This is due to a number of factors, including increased awareness and changes in the construction of buildings. Energy concerns have caused buildings that are nearly completed to be sealed to prevent air from leaking into the building and to keep heat and air conditioning inside. This has resulted in "tight buildings" which tend to retain any contamination that is introduced into the building. The resulting "Sick Building Syndrome" may cause health problems among employees in the building, or may result in decreased productivity due to poor working conditions.

Designing the HVAC System

The most important element for preventing indoor air problems is proper ventilation. Ventilation systems provide remove contaminants from areas where they would otherwise accumulate. It also provides fresh, clean air for employees working in the area.

If you have the luxury of being part of the design team for a building, make sure that particular attention is paid to the HVAC system. The flow of air through an area should cover the entire area. Too often, the inlet and outlet for fresh air are located next to each other and form a closed loop that does not allow air to circulate throughout an area. Ideally, an inlet should be located near the ceiling on one side of an area, and the outlet should be located near the floor opposite the inlet. This provides maximum circulation through the area. Of course, a single inlet will not be sufficient for large areas which will require more circulation than can be provided by a single inlet.

Also, pay attention to the amount and type of make-up air that will be provided to the system. At a minimum, at least 10 cubic feet of make-up air per minute should be provided per person in an area. The make-up air should contain at least ten percent fresh, outdoor air. Consider energy use when deciding the amount of fresh air to be introduced into the system. The best system would provide 100% fresh air; however, this would lead to astronomical heating and air conditioning costs.

You should also have an idea of the types of processes that will be carried out within the facility. An area that will be used for chemical processes may need a separate HVAC system from the rest of the building.

Once the system is installed, it should be inspected. Look for areas that may accumulate water or other contaminants. Standing water in HVAC systems allows bacteria to grow. Be sure that all fans within the system are moving air in the proper direction. Switching the wires may result in the system actually increasing the contamination in your facility, rather than providing employees with clean air.

Maintaining the HVAC System

Assuming that your HVAC system is properly designed, maintenance of the system is the most important method of preventing indoor air problems. The manufacturer should provide general preventative maintenance procedures to follow with your particular system. At least annually, you should inspect the system and conduct proper cleaning. Areas which generate large amounts of dust may need duct work cleaned more often.

If you know the capacity of your system and the flow rates that should be present at each of the outlets and inlets, use a portable anemometer to measure the flow rates. You can rent anemometers from most environmental equipment rental companies. If the flow rates do not match the designed flow rates, there may be a blockage in the system, or the fan may not be operating properly. Be sure to keep records of all flow rates and maintenance procedures.

Choosing Appropriate Furnishings and Building Materials

A number of materials that are part of the building or that are brought into the building may result in indoor air quality problems. The most common building materials are particle board and adhesives. Particle board may emit formaldehyde vapors into the building. This will result in employees complaining of burning eyes, or respiratory irritation. Other materials may also emit formaldehyde and other organic vapors. They include furnishings and carpeting. To determine if furnishings or building materials are causing indoor air problems, first determine if any new materials have entered your building. If you are unable to locate the problem, an environmental company with indoor air capabilities can provide sampling to determine the nature of the contaminants present.

If the problems you are experiencing are due to furnishings, remove the furnishings to a remote area until they have been allowed sufficient time to "air out." Generally, new furnishings will eventually cease to emit vapors. If the materials causing the problem are too large to be removed from the building, you may be able to air out the entire building. This may be as simple as opening the fresh air intakes on your HVAC system to 100 percent for a weekend. Increasing the temperature within the building may also help to drive off vapor quicker than they normally be emitted from the material. If this does not work, you may have to add high-capacity fans in the area to increase the volume of air moving through the building. If none of this work, you may need to remove the material in question entirely from the building.

The best means of preventing this type of problem is to evaluate materials as they come into the building. If there is a question, contact the manufacturer or other persons who have bought the product in the past.

Solving Occupant-Generated Problems

The most common occupant-generated problem is the accumulation of carbon dioxide in the building. Carbon dioxide is the product of persons breathing. This is the result of poor make-up air and circulation within the building.

Carbon dioxide is not a toxic substance, but it may replace oxygen in the area, resulting in an oxygen deficient atmosphere. The oxygen levels do not normally reach life threatening proportions, but they may result in decreased alertness. Employees may complaining of fatigue and general malaise. Generally, the problem becomes increasingly worse throughout the day, as more carbon dioxide is generated by occupants.

Other occupant-generated problem include smoking, use of nail polish, and perfumes. Many persons are extremely sensitive or allergic to these materials, and may be effected if the materials are not removed from the indoor air.

You can correct oxygen deficiencies by increasing the flow and circulation provided by your HVAC system. You can alleviate other problems by altering the HVAC system or by using administrative controls, such as eliminating smoking from your facility. Administrative controls require a commitment from management.

Solving Outdoor Health Problems

While introducing fresh air into the building is important to the health of the building, the air that is introduced into the system may not always be clean. One of the most common problems is locating the fresh air intake near an obvious source of contamination such as a loading dock, or process vent. This is something that designers do not readily recognize, if they have never visited the sight. One case occurred where the fresh air intake for a building was located directly across from a dyeing process vent. This resulted in solvents being introduced at high enough concentrations to cause employees to becomes ill.

If you are introducing outside contaminants into your building, you need to relocate the fresh air intake. This may result in adding an extension to the intake so that air is being introduced from outside the contaminated area. This is the most cost effective means to remedy the problem. If this can not be done, you may need to completely relocate and redesign your HVAC system.

6

CFC and HCFC Refrigeration Equipment Maintenance

Evans J. Lizardos, P.E.

Because of the ozone depletion effects of chlorofluorocarbons (CFCs), the U.S. 1990 Clean Air Act calls for progressive reductions in CFC and HCFC (hydrochlorofluorocarbon) production, with zero production of CFCs by the year 1996 and HCFCs by the year 2030 (2003 and 2010 for certain HCFCs).

CURRENT STEPS TOWARD REDUCTIONS IN CFCs AND HCFCs

As of January 1, 1993, all of the major manufacturers have committed to a curtailment of CFC production by January 1, 1996. Also, all of the major manufacturers of centrifugal chiller equipment have announced that they will no longer use CFCs in their equipment.

At the November, 1992 United Nations meeting in Copenhagen, Denmark, revisions to the Montreal Protocol resulted in the first phase out schedule for HCFCs (see Table 6-1).

In addition, since July 1, 1992, the U.S. Clean Air Act has prohibited the intentional venting of CFC and HCFC refrigerants during equipment operation, maintenance, repairs, or disposal.

Both companies and technicians can be held accountable for intentional CFC/HCFC venting violations. Both eyewitness testimony and records examinations will constitute proof of intentional venting of ozone-depleting refrigerants. However, the following are not in violation of the new rules: accidental refrigera-

Jan. 1	3.1% of ODP* of 1989 CFC production plus total ODP* 1989 HCFC production
2004	65% of cap
2010	35% of cap
2015	10% of cap
2020	0.5% of cap
2030	Total phaseout

*ODP = Ozone Depletion Potential

Table 6-1. HCFC Phaseout as of January 1, 1996

tion leaks; topping-off refrigerant in a system; leak detection refrigerant gases; and holding charges for equipment being shipped.

Adhering to this new legislation may mean several changes to your maintenance practices, depending on the type of equipment and refrigerant used. New requirements for your equipment may include equipment purge control, refrigerant storage and recycling equipment, personnel certification, equipment venting, and relief valves.

Maintenance managers must choose between continued CFC refrigeration equipment service or the purchase of new non-CFC refrigeration equipment. For equipment with a continued useful service life, the cost outlays for new maintenance procedures such as refrigerant recycling and certification may be significantly less than equipment replacement or equipment retrofits with CFC substitutes. Equipment retrofits can adversely affect system capacity and efficiency.

EQUIPMENT MAINTENANCE GUIDELINES FOR LARGE TONNAGE AND SMALL TONNAGE REFRIGERATION SYSTEMS

Following is a discussion of equipment maintenance guidelines for both large tonnage and smaller tonnage refrigeration systems.

Identifying Large Tonnage Machines

Large tonnage machines include the following:

- low-pressure CFC-11 and HCFC-123 refrigeration equipment
- medium-pressure CFC-12 refrigeration equipment
- high-pressure HCFC-22 refrigeration equipment

- medium-pressure HFC-134a refrigerant equipment; this is the new refrigerant that has a zero ozone depletion potential and, though the following maintenance requirements may be applicable for HFC-134a, they are not mandated.

Maintenance Requirements for Large Tonnage Machines

The following devices are required for maintenance of large tonnage machines:

Purge Control

Large chillers operating at lower than atmospheric pressures must be periodically purged of air which seeps into the system. While purging, uncondensed refrigerant will be released along with the purged air. Many of the older purge devices allowed 3 to 8 pounds of refrigerant to escape per pound of air purged.

Extremely efficient purge devices are built today (99 percent efficiency) and should be utilized to minimize refrigerant loss and escape to the atmosphere. Also, purge recycle devices are available which capture, store and filter refrigerants released during system purge.

Purge devices would not be required for medium-pressure and high-pressure systems as they operate above atmospheric pressure.

Refrigerant Pump-Out

Large tonnage machines holding refrigerant charges of several pounds must be equipped with pump-out units to store the refrigerant charge during system maintenance. Pump-out evacuation levels in mm. Hg. are specified for low, medium and high-pressure systems in the EPA recovery and recycling rule.

Low pressure machines require permanent refrigerant pump out units, mainly due to the periods of time when they are not in use, specifically, when the chillers are secured during winter and the machine stands idle for many months. Under this condition, if the chiller is not pumped out, it could collect incondensibles during its idle period.

At this time, some manufacturers have equipped their machines with devices that allow the pressure to remain above atmospheric and prevent this unwanted occurrence. In either case, it is important to take precautions on low pressure machines that are idle and have the possibility of going into subatmospheric conditions.

Relief Pressure Devices

There are two basic types of relief pressure devices: spring relief valves and rupture discs. Spring relief valves are preferred for refrigerant containment because the valve recloses once the pressure returns to normal, retaining much of the refrigerant charge. In contrast, rupture discs release the full charge once they are engaged. Large tonnage machines commonly employ rupture discs. Applications using rupture discs should be considered for replacement with relief valves.

Figures 6-1 and Figure 6-2 illustrate methods for piping in series the rupture disc and relief valve. These arrangements allow for the rupture disc to be in series

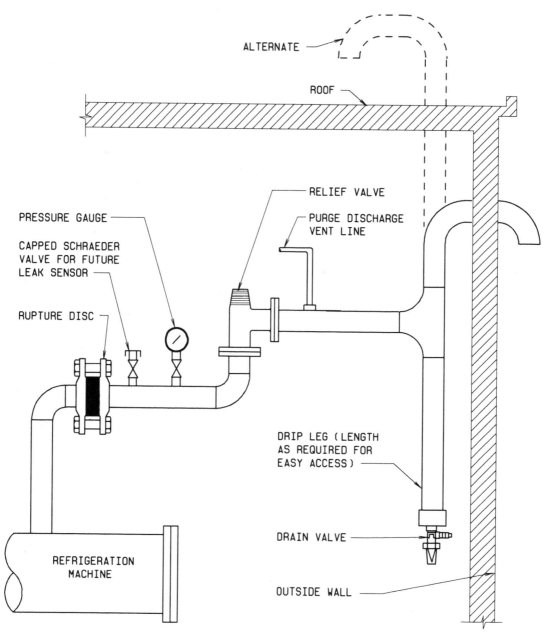

Figure 6-1. Schematic of Rupture Disc and Relief Valve Piping Arrangement

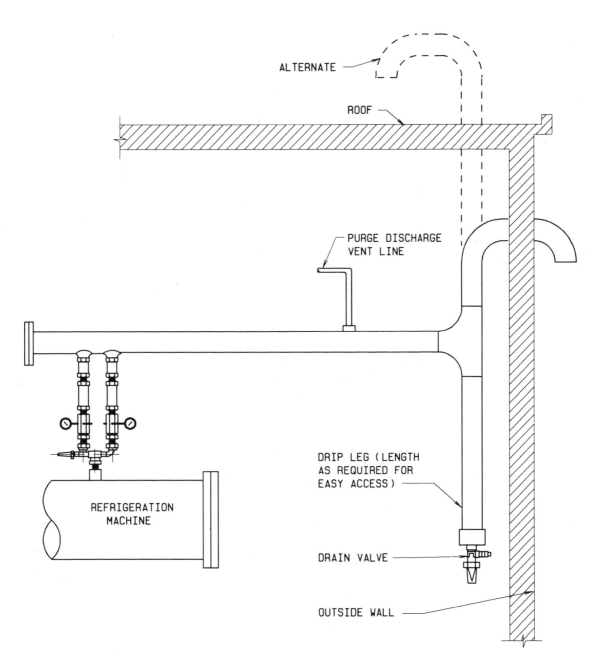

Figure 6-2. Schematic of Rupture Discs on a Dual Relief Valve
Piping Arrangement

with the relief valve. The relief valve can be used as a detecting means if the disc ruptures. Refrigerant loss is measured by the system pressure gauge which indicates how much refrigerant has been released through the rupture disc.

Equipment Venting

In the past, rupture discs and relief valves have been vented to the outside. Although this practice introduces CFCs/HCFCs to the environment, the relief valve operation is considered an emergency feature and outdoor venting of pressure relief devices is still permitted.

Mechanical Room Ventilation

Mechanical room ventilation has been revised by the new "ANSI/ASHRAE 15a -1991 Safety Code for Mechanical Refrigeration" and is based on the new "ANSI/ASHRAE 34 -1992 Number Designation and Safety Classification of Refrigerants." Essentially, either an oxygen deprivation sensor set at levels of 19.5 percent volume or a refrigerant sensor set no higher than the refrigerant "Threshold Limit Value (TLV)" is required which will alarm and automatically start exhausting the space. As an example, the following chart illustrates which sensor types are required for some commonly used or proposed new refrigerants (see Table 6-2).

Refrigerant No.	Safety Group	Sensor Type
11	A1	Oxygen Deprivation
12	A1	Oxygen Deprivation
22	A1	Oxygen Deprivation
113	A1	Oxygen Deprivation
114	A1	Oxygen Deprivation
115	A1	Oxygen Deprivation
123	B1	Refrigerant Sensor
134a	A1	Oxygen Deprivation
500	A1	Oxygen Deprivation
502	A1	Oxygen Deprivation
717 ammonia	B1	Refrigerant Sensor

Table 6-2. Sensors to be Used with Various Refrigerants

Personnel and Equipment Certification

As of April, 1993, the portion of the rule covering personnel certification by the Environmental Protection Agency (EPA) requires mandatory certification.

The personnel certification program has four parts, based on equipment type:

- Small appliances with less than 1 lb of refrigerant;
- High-pressure;
- Low-pressure equipment;
- Universal certification covering all types.

The equipment certification standards are based on depth of vacuum (see Table 6-3).

System Type	(Recovery-Recycler Made Before 11/15/93)	(Recovery-Recycler Made After 11/15/93)
High pressure below 200 lb charge	4	10
High pressure over 200 lb charge	4	15
Very high pressure	0	0
Intermediate pressure	25	25
Low pressure	25	25 mm Hg absolute

Table 6-3. Evacuation Levels for Recovery-Recycling Machines (Except Small Appliances) in Hg Vacuum

Identifying Smaller Tonnage Machines

Smaller tonnage machines include the following:

- reciprocating compressors
- scroll compressors
- screw compressors
- employing CFC-12, CFC-502 and HCFC-22
- employing HFC-134a (HFC-134a is a new refrigerant which has a zero ozone depletion potential. and though the following maintenance requirements may be applicable for HFC-134a, they are not mandated.

Maintenance Requirements for Smaller Tonnage Machines

The following devices are required for maintenance of smaller tonnage machines.

Purge

In general, smaller tonnage machines operate at above atmospheric pressures and so purge devices are not required.

Recovery/Recycling Units

CFC and HCFC recovery and recycling equipment is required to hold charges of CFCs/HCFCs during servicing, and to filter and clean the charge for reuse.

Recovery and recycling equipment can extend the service life of CFC/HCFC refrigeration equipment well past the end of CFC production. Also, recovery and recycling equipment provides economy as the price of CFCs escalates with the decrease in production.

Many types of portable recovery and recycling equipment are available. Some less costly units make use of the refrigeration equipment compressor to reload the stored charge. Other equipment contains built-in filters, dryers and compressors. Recovery and recycling equipment should be rated and certified according to type of refrigeration equipment and evacuation levels (mm Hg) required by the EPA rule previously discussed.

Relief Pressure Devices

Most smaller tonnage refrigeration equipment makes use of rupture discs for emergency pressure relief. On small machines, such as household units of 1-2 lb. charge, rupture discs are permitted because spring relief valves are not feasible or possible.

Equipment Venting

As discussed for large tonnage refrigeration equipment, outside venting of relief pressure devices on small tonnage equipment is permitted because the discs are an emergency condition only. Small systems may be vented to room air. For larger systems confined in small spaces, venting to the outside may be required so that released CFC/HCFC do not deplete room oxygen. (Refer to the ANSI/ASHRAE 15a-1991).

Certification

Certification for smaller tonnage machines of both personnel and equipment applies as discussed for large tonnage machines.

Index